Mastering Essential Math Skills

Decimals and Percents

Richard W. Fisher

Decimals and Percents - Item #406
ISBN 10: 0-9666211-6-6 • ISBN 13: 978-0-9666211-6-7

1st printing 2008; 2nd printing 2010; 3rd printing 2012; 4th printing 2014

Notes to the Teacher or Parent

What sets this book apart from other books is its approach. It is not just a math book, but a system of teaching math. Each daily lesson contains four key parts: **Review Exercises**, **Helpful Hints**, **New Material**, and **Problem Solving**. Teachers have flexibility in introducing new topics, but the book provides them with the necessary structure and guidance. The teacher can rest assured that essential math skills in this book are being systematically learned.

This easy-to-follow program requires only fifteen or twenty minutes of instruction per day. Each lesson is concise and self-contained. The daily exercises help students to not only master math skills, but also maintain and reinforce those skills through consistent review - something that is missing in most math programs. Skills learned in this book apply to all areas of the curriculum, and consistent review is built into each daily lesson. Teachers and parents will also be pleased to note that the lessons are quite easy to correct.

This book is based on a system of teaching that was developed by a math instructor over a thirty-year period. This system has produced dramatic results for students. The program quickly motivates students and creates confidence and excitement that leads naturally to success.

Please read the following "How to Use This Book" section and let this program help you to produce dramatic results with your math students.

How to Use This Book

This book is best used on a daily basis. The first lesson should be carefully gone over with students to introduce them to the program and familiarize them with the format. It is hoped that the program will help your students to develop an enthusiasm and passion for math that will stay with them throughout their education.

As you go through these lessons every day, you will soon begin to see growth in the student's confidence, enthusiasm, and skill level. The students will maintain their mastery through the daily review.

Step 1

The students are to complete the review exercises, showing all their work. After completing the problems, it is important for the teacher or parent to go over this section with the students to ensure understanding.

Step 2

Next comes the new material. Use the "Helpful Hints" section to help introduce the new material. Be sure to point out that it is often helpful to come back to this section as the students work independently. This section often has examples that are very helpful to the students.

Step 3

It is highly important for the teacher to work through the two sample problems with the students before they begin to work independently. Working these problems together will ensure that the students understand the topic, and prevent a lot of unnecessary frustration. The two sample problems will get the students off to a good start and will instill confidence as the students begin to work independently.

Step 4

Each lesson has problem solving as the last section of the page. It is recommended that the teacher go through this section, discussing key words and phrases, and also possible strategies. Problem solving is neglected in many math programs, and just a little work each day can produce dramatic results.

Step 5

Solutions are located in the back of the book. Teachers may correct the exercises if they wish, or have the students correct the work themselves.

Table of Contents

Review Exercises

Note to the students and teachers: This section will include daily review from all topics covered in this book. Here are some simple problems to get started.

1. 36 + 17 + 44 =

2. 521 - 306 =

3. 715 - 264

4. 47 + 348 + 225

5. Find the sum of 25, 36, and 48.

6. Find the difference between 700 and 76.

Helpful Hints

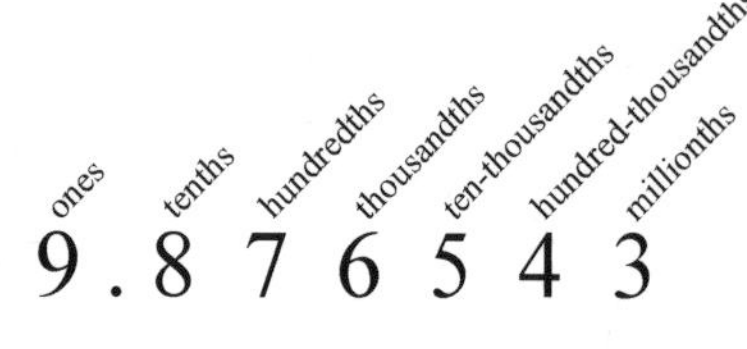

To read decimals, first read the whole number. Next, read the decimal point as "and." Next, read the number after the decimal point and its place value.

Examples:

3.16 = three and sixteen hundredths
14.011 = fourteen and eleven thousandths
0.69 = sixty-nine hundredths

Write the following in words.

S1. 3.7

S2. 12.019

1. 0.87

2. 5.006

3. 115.7

4. 78.07

5. 6.3912

6. 0.085

7. 7.36

8. 9.002

9. 0.61

10. 2.333

1.
2.
3.
4.
5.
6.
7.
8.
9.
10.
Score

Problem Solving

Juan has 709 dollars and Al has 529 dollars. How many more dollars does Juan have than Al?

Review Exercises

1.
```
  427
  816
   23
+ 142
```

2. Find the difference between 1,726 and 977.

3. 601 - 78 =

4. 37 + 42 + 36 =

5.
```
  500
- 276
```

6.
```
  315
x   4
```

Helpful Hints

Use what you have learned to solve the following problems.
* Read the decimal point as "and".
* Put hyphens in numbers between 20 and 99 when necessary.

Write the following in words.

S1. 3.006

S2. 0.0176

1. 0.8

2. 3.0005

3. 76.8

4. 7.008

5. 5.138

6. 0.015

7. 5.82

8. 4.03

9. 0.86

10. 4.224

1.
2.
3.
4.
5.
6.
7.
8.
9.
10.
Score

Problem Solving

Three classes at Hoover School have enrollments of 37, 48, and 40. What is the total enrollment of the three classes?

Review Exercises

1. 336
   19
   + 424

2. Write 2.007 in words.

3. 7,125
   - 743

4. Write 42.016 in words.

5. 37 + 16 + 274 =

6. Write 0.019 in words.

Helpful Hints

When reading, remember "and" means decimal point. The fraction part of a decimal ends in "th" or "ths."
Be careful about placeholders.

Example:
Four and eight tenths = 4.8
Two hundred one and six hundredths = 201.06
One hundred four ten-thousandths = .0104

Write each of the following as a decimal. Use the chart at the bottom to help.

S1. Five and three hundredths.

S2. Four hundred thirty-six and eleven hundredths.

1. Seven and four tenths.
2. Twenty-two and fifteen thousandths.
3. Three hundred fifty-two ten-thousandths.
4. Seventy-four and forty-three thousandths.
5. Five hundred-thousandths.
6. Sixteen millionths.
7. Nine and forty-five thousandths.
8. Twenty and thirty-three ten-thousandths.
9. Eighty-six and nine tenths.
10. Eighty-six and nine millionths.

ones tenths hundredths thousandths ten-thousandths hundred-thousandths millionths
9 . 8 7 6 5 4 3

1.
2.
3.
4.
5.
6.
7.
8.
9.
10.
Score

Problem Solving

Crayons come in boxes of 24. How many crayons are there in fifteen boxes?

Review Exercises

1. Write 2.09 in words.

2. Write 2.009 in words.

3. Two and four hundredths equals what decimal?

4. 761 - 79 =

5. Fifteen thousandths equals what decimal?

6.
```
  7,647
    362
+ 5,173
```

Helpful Hints

Use what you have learned to solve the following problems.
* "and" means decimal point.
* The fraction part of a decimal ends in "th" or "ths".
* Be careful about placeholders.

Write each of the following as a decimal. Use the chart at the bottom to help.

S1. Seven and sixty-two ten-thousandths..

S2. Two thousand nine hundred-thousandths.

1. Eight and nine hundredths.

2. Twelve and forty-one hundred-thousandths.

3. Forty-nine ten-thousandths.

4. Ninety-seven and five hundred thirteen millionths.

5. Forty-eight thousandths.

6. Fifty-two and eight tenths.

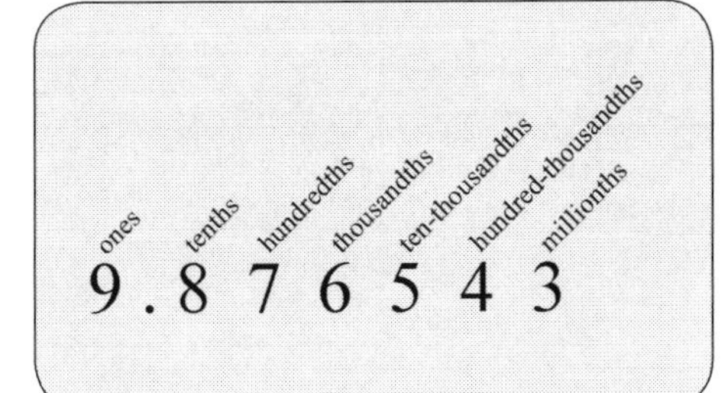

7. Five and four hundred ninety-six thousandths.

8. Three and five thousandths.

9. Twelve and thirty-three hundred-thousandths.

10. One hundred sixteen and five hundredths.

1.
2.
3.
4.
5.
6.
7.
8.
9.
10.
Score

Problem Solving

360 students are placed in fifteen equally-sized classes.
How many students are in each class?

Review Exercises

1. Write 2.07 in words.
2. Write 7.017 in words.
3. Write seven and six thousandths as a decimal?
4. Write thirty-two millionths as a decimal?
5. Write 0.0017 in words.
6. Write five and eleven ten-thousandths as a decimal?

Helpful Hints

When changing mixed numerals to decimals, remember to put a decimal after the whole number.

Examples: $3\frac{3}{10} = 3.3$ $\quad \frac{16}{10,000} = .0016$ $\quad 42\frac{9}{10,000} = 42.0009$ $\quad 65\frac{12}{100,000} = 65.00012$

Write each of the following as a decimal. Use the chart at the bottom to help.

S1. $5\frac{6}{10}$

S2. $8\frac{9}{1,000}$

1. $21\frac{16}{100}$
2. $\frac{16}{100}$
3. $14\frac{17}{1,000}$
4. $119\frac{16}{100,000}$
5. $\frac{21}{10,000}$
6. $3\frac{196}{100,000}$
7. $4\frac{32}{1,000}$
8. $3\frac{324}{1,000}$
9. $4\frac{17}{1,000,000}$
10. $\frac{19}{10,000}$

ones, tenths, hundredths, thousandths, ten-thousandths, hundred-thousandths, millionths

9 . 8 7 6 5 4 3

1.
2.
3.
4.
5.
6.
7.
8.
9.
10.
Score

Problem Solving

A plane traveled 3,150 miles in seven hours. What was its average speed per hour?

Review Exercises

1. Write ten and fourteen thousandths as a decimal.

2. Write 2.09 in words.

3. Write sixty-five thousandths as a decimal.

4. Write ten and fifteen hundred-thousandths as a decimal.

5. 724 - 617 =

6. Find the difference between 972 and 408.

Helpful Hints

Use what you have learned to solve the following problems.
* Remember to put a decimal point after the whole number.
* Be careful to use placeholders when necessary.

Write each of the following as a decimal. Use the chart at the bottom to help.

S1. $9\frac{17}{1,000}$

S2. $9\frac{17}{100,000}$

1. $42\frac{196}{1,000}$

2. $\frac{72}{1,000}$

3. $48\frac{8}{1,000}$

4. $16\frac{195}{100,000}$

5. $\frac{16}{1,000}$

6. $16\frac{119}{1,000,000}$

7. $4\frac{38}{10,000}$

8. $3\frac{176}{10,000}$

9. $\frac{71}{1,000}$

10. $6\frac{53}{10,000}$

ones		tenths	hundredths	thousandths	ten-thousandths	hundred-thousandths	millionths
9	.	8	7	6	5	4	3

1.
2.
3.
4.
5.
6.
7.
8.
9.
10.

Score

Problem Solving

If a car can travel 32 miles for each gallon of gas that it uses, how far can the car travel using 12 gallons of gas?

Review Exercises

1. Write $9\frac{7}{100}$ as a decimal.
2. Write $1\frac{135}{100{,}000}$ as a decimal.
3. Write seventeen hundred-thousandths as a decimal.
4. Write 0.021 in words.
5. Write 3.19 in words.
6. Write six and thirteen ten-thousandths as a decimal.

Helpful Hints

Decimals can easily be changed to mixed numbers and fractions. Remember that the whole number is to the left of the decimal.

Examples: $2.6 = 2\frac{6}{10}$ $.210 = \frac{210}{1{,}000}$ $3.007 = 3\frac{7}{1{,}000}$ $1.0019 = 1\frac{19}{10{,}000}$

Change the following to a mixed numeral or fraction. Use the chart at the bottom for help.

S1. 3.05 S2. 16.017

1. 45.0019
2. .00005
3. 7.000016
4. 7.196
5. 79.6
6. .07632
7. 14.00007
8. 16.024
9. 17.000145
10. .00096

1.
2.
3.
4.
5.
6.
7.
8.
9.
10.

Score

Problem Solving

A worker earned 750 dollars. He spent 300 dollars for his rent and 176 dollars for his car payment. How much of his earnings were left?

Review Exercises

1. Write eleven and six hundredths as a decimal.

2. Write six thousandths as a fraction.

3. Write $\frac{19}{1,000}$ as a decimal.

4. Write 7.006 as a mixed numeral.

5. Write six and three tenths as a mixed numeral.

6. Write seventy-two thousandths as a decimal.

Helpful Hints

Use what you have learned to solve the following problems.

* Remember, the whole number is to the left of the decimal.

Change the following to a mixed numeral or fraction.
Use the chart at the bottom for help.

S1. .0016

S2. 9.00125

1. 7.00009

2. .016

3. 7.29

4. 6.00002

5. 87.3

6. 5.0072

7. 15.000006

8. 42.1

9. 163.0137

10. .11234

ones tenths hundredths thousandths ten-thousandths hundred-thousandths millionths

9 . 8 7 6 5 4 3

1.
2.
3.
4.
5.
6.
7.
8.
9.
10.
Score

Problem Solving

June wants to send our invitations to 86 people. The invitations come in packages of 15. How many packages must she buy. How many cards will be left over?

Review Exercises

1. Write 7.0016 as a mixed numeral.

2. Write $7\frac{15}{1,000}$ as a decimal.

3. Write five and six tenths as a decimal.

4. Write 6.013 in words.

5. Write $3\frac{7}{1,000}$ in words.

6. Write 72.0012 as a mixed numeral.

Helpful Hints

Zeroes can be put to the right of a decimal without changing the value. This helps when comparing the value of decimals.

< means less than

> means greater than

Example:
Compare 4.3 and 4.28
4.3 = 4.30 so 4.3 > 4.28

Place > or < to compare each pair of decimals. Use the chart at the bottom for help.

S1.	7.46 ☐ 7.5	S2.	.88 ☐ .879
1.	3.099 ☐ 3.1	2.	7.52 ☐ 7.396
3.	6.22 ☐ 6.31	4.	1.7 ☐ 1.69
5.	2.21 ☐ 1.99	6.	.4 ☐ .71
7.	2.29 ☐ 2.4	8.	3.65 ☐ 3.589
		9.	9.09 ☐ 9.2
		10.	8.199 ☐ 8.2

ones, tenths, hundredths, thousandths, ten-thousandths, hundred-thousandths, millionths

9 . 8 7 6 5 4 3

1.
2.
3.
4.
5.
6.
7.
8.
9.
10.
Score

Problem Solving

A car traveled 240 miles. For each 30 miles traveled, the car used one gallon of gas. How many gallons of gas were used on the trip? If gas cost 3 dollars per gallon, how much did the trip cost?

Review Exercises

1. Write .0129 as a fraction.

2. Write $7\frac{92}{1,000}$ in words.

3. Write .0135 in words.

4. Write seventeen and six thousandths as a decimal.

5. Change $15\frac{71}{10,000}$ to a decimal.

6. Change 3.096 to a mixed numeral.

Helpful Hints

Use what you have learned to solve the following problems.
* Add zeroes to help compare the values.

Place > or < to compare each pair of decimals. Use the chart at the bottom for help.

S1. 8.6 ☐ 8.72

S2. .799 ☐ .81

1. 3.196 ☐ 3.2

2. 2.423 ☐ 2.398

3. 6.12 ☐ 6.115

4. 3.7 ☐ 3.68

5. .9762 ☐ 1.001

6. .4 ☐ .53

7. 2.41 ☐ 2.296

8. 4.63 ☐ 4.596

9. .9 ☐ .875

10. 3.09 ☐ 3.112

ones	tenths	hundredths	thousandths	ten-thousandths	hundred-thousandths	millionths
9.	8	7	6	5	4	3

1.
2.
3.
4.
5.
6.
7.
8.
9.
10.
Score

Problem Solving

Leo, Maria, Pedro, and Jill together earned 724 dollars. If they wanted to share the money equally, how much will each one receive?

Review Exercises

1. Change 52.623 to a mixed numeral.

2. Write $75\frac{9}{1,000}$ as a decimal.

3. $\begin{array}{r} 624 \\ \times\ 7 \\ \hline \end{array}$

4. Write fifty and nineteen thousandths as a decimal.

5. Write 6.23 in words.

6. 7,051 - 267 =

Helpful Hints

To add decimals, line up the decimal points and add as you would whole numbers. Write the decimal points in the answer. Zeroes may be placed to the right of the decimal.

Example: Add 3.16 + 2.4 + 12

$\begin{array}{r} 3.16 \\ 2.40 \\ +\ 12.00 \\ \hline 17.56 \end{array}$

S1. $\begin{array}{r} 3.16 \\ 12.4 \\ +\ 3.26 \\ \hline \end{array}$

S2. 3.92 + 4.6 + .32 =

1. 32.16 + 1.7 + 7.493 =

2. 7.341 + 6.49 + .6 =

3. $\begin{array}{r} 7.64 \\ 19.633 \\ +\ 2.4 \\ \hline \end{array}$

4. .37 + .6 + .73 =

5. 9.64 + 7 + 1.92 + .7 =

6. 72.163 + 11.4 + 63.42 =

7. .7 + .6 + .4 =

8. 17.33 + 6.994 + .72 =

9. $\begin{array}{r} 7.642 \\ 17.63 \\ 2.143 \\ +\ 14.64 \\ \hline \end{array}$

10. 19.2 + 7.63 + 4.26 =

1.
2.
3.
4.
5.
6.
7.
8.
9.
10.
Score

Problem Solving

In January it rained 4.76 inches, in February, 6.43 inches, and in March, 7.43 inches. What was the total amount of rainfall for the three months?

Review Exercises

1. Write 7.19 in words.

2. $$\begin{array}{r} 7.63 \\ 4.2 \\ +\ 2.673 \\ \hline \end{array}$$

3. Write $7\frac{5}{10,000}$ as a decimal.

4. Write 7.0632 as a mixed numeral.

5. Write .019 in words.

6. .7 + .4 + .8 =

Helpful Hints

Use what you have learned to solve the following problems.

* Zeros may be placed to the right of the decimal.
* Write the decimal point in the answer.

S1. $$\begin{array}{r} 4.6 \\ 13.24 \\ +\ 7.89 \\ \hline \end{array}$$

S2. 3.67 + 9.3 + .61 =

1. 41.23 + 6.973 + 1.9 =

2. 5.912 + 5 + 7.63 =

3. $$\begin{array}{r} 19.62 \\ 7.426 \\ +\ 7.93 \\ \hline \end{array}$$

4. .36 + .79 + .6 =

5. 92.5 + 12 + 2.16 + .7 =

6. $$\begin{array}{r} 73.197 \\ 6.72 \\ 17.109 \\ +\ 6.28 \\ \hline \end{array}$$

7. .9 + .6 + .8 =

8. 76.76 + 8.765 + .89 =

9. $$\begin{array}{r} 9.726 \\ 14.276 \\ 7.93 \\ +\ 24.67 \\ \hline \end{array}$$

10. 17.6 + 3.68 + 4.27 =

1.
2.
3.
4.
5.
6.
7.
8.
9.
10.
Score

Problem Solving

Mr. Otis purchased milk for $3.15 and bread for $1.85. If he paid with a 20 dollar bill, what was the amount of change that he should receive?

Review Exercises

1.
$$\begin{array}{r} 7.24 \\ 3.6 \\ +\ 8.717 \\ \hline \end{array}$$

2. $4.8 + 6 + 3.26 =$

3. Write $7\frac{125}{10{,}000}$ as a decimal.

4. Write 6.012 as a mixed numeral.

5. Write $2\frac{7}{100}$ in words.

6. Write 7.005 in words.

Helpful Hints

To subtract decimals, line up the decimal points and subtract as you would whole numbers. Write the decimal points in the answer. Zeroes may be placed to the right of the decimal.

Examples: 3.2 - 1.66 = 7 - 1.63 =

$$\begin{array}{r} \overset{2\ 11\ 1}{\not{3}.\not{2}0} \\ -\ 1.66 \\ \hline 1.54 \end{array} \qquad \begin{array}{r} \overset{6\ 9\ 1}{\not{7}.\not{0}0} \\ -\ 1.63 \\ \hline 5.37 \end{array}$$

S1.
$$\begin{array}{r} 17.2 \\ -\ 3.36 \\ \hline \end{array}$$

S2. $15.1 - 7.62 =$

1.
$$\begin{array}{r} 7.32 \\ -\ 1.426 \\ \hline \end{array}$$

2.
$$\begin{array}{r} 3.962 \\ -\ 1.669 \\ \hline \end{array}$$

3. $2.72 - 1.56 =$

4. $27.93 - 16.8 =$

5. $.72 - .667 =$

6.
$$\begin{array}{r} 6.137 \\ -\ 2.1793 \\ \hline \end{array}$$

7. $3 - .627 =$

8. $7.14 - 3.456 =$

9. $75.6 - 66.972 =$

10. $43.21 - 16.445 =$

1.
2.
3.
4.
5.
6.
7.
8.
9.
10.
Score

Problem Solving

The normal temperature is 98.6°. If Gwen's temperature is 101.3°, how much above normal is her temperature?

Review Exercises

1. Which has the largest value, 2.6 or 2.599?

2. 6.23 + 2.7 + 3.24 =

3. Write $\frac{72}{100}$ in words.

4. Write $\frac{72}{10,000}$ in words.

5. Write 75.0006 as a mixed numeral.

6. 15.8 + 75 + 6.23 =

Helpful Hints

Use what you have learned to solve the following problems.
* Line up the decimal points.
* Zeroes may be added to the right of the decimal.
* Write the decimal point in the answer.

S1. 75.3 - 19.68 =

S2. 75 - 1.96 =

1. $\begin{array}{r} 6.35 \\ -\ 2.347 \\ \hline \end{array}$

2. $\begin{array}{r} 7.547 \\ -\ .36 \\ \hline \end{array}$

3. 7.15 - 3.672 =

4. 37.3 - .96 =

5. .89 - .697 =

6. $\begin{array}{r} 7.136 \\ -\ 2.2476 \\ \hline \end{array}$

7. 5 - .964 =

8. 95.1 - 7.124 =

9. 6.2 - 3.914 =

10. 85.3 - 27.965 =

1.

2.

3.

4.

5.

6.

7.

8.

9.

10.

Score

Problem Solving

Mrs. Roberts had $210.25 in her savings account. On Monday she withdrew $56.75 and on Friday she made a deposit of $125.50. How much does she now have in her savings account?

Review Exercises

1. $$\begin{array}{r} 7.63 \\ 9.4 \\ +\ .76 \\ \hline \end{array}$$

2. $$\begin{array}{r} 7.6 \\ -\ 2.542 \\ \hline \end{array}$$

3. 9.6 + 5 + 2.76 =

4. 15 - 2.78 =

5. Write $\frac{75}{1,000}$ as a decimal.

6. Write 2.058 in words.

Helpful Hints

Use what you have learned to solve the following problems.
* Line up the decimals.
* Put decimals in the answer.
* Zeroes may be added to the right of the decimal.

S1. $$\begin{array}{r} 3.61 \\ 14.4 \\ +\ .37 \\ \hline \end{array}$$

S2. $$\begin{array}{r} 7.16 \\ -\ 3.473 \\ \hline \end{array}$$

1. $$\begin{array}{r} 7.61 \\ 8.92 \\ +\ 7.634 \\ \hline \end{array}$$

2. $$\begin{array}{r} 7.6 \\ -\ 1.43 \\ \hline \end{array}$$

3. 4.63 + 5.7 + 6.24 =

4. 17.2 - 8.96 =

5. 15 - 12.92 =

6. 6.93 + 5 + 7.63 =

7. .9 + .7 + .6 =

8. 7.16 - 2.673 =

9. 27.16 - 16.764 =

10. 7.73 + 2.6 + .37 + 15 =

Answers
1.
2.
3.
4.
5.
6.
7.
8.
9.
10.
Score

Problem Solving

In May Roberto weighed 145.2 pounds. In July he weighed 139.7 pounds. How much more did he weigh in May than in July?

Review Exercises

1. Write $\frac{7}{1,000,000}$ as a decimal.

2. Write $\frac{7}{1,000,000}$ in words.

3. $\begin{array}{r} 62.7 \\ 3.9 \\ 5. \\ +\ 7.62 \\ \hline \end{array}$

4. Write six and twenty-two ten-thousandths as a decimal.

5. 7 - 2.36 =

6. $\begin{array}{r} 5.1 \\ -\ 2.765 \\ \hline \end{array}$

Helpful Hints

Use what you have learned to solve the following problems.

S1. $\begin{array}{r} 42.6 \\ .39 \\ +\ 6.427 \\ \hline \end{array}$

S2. $\begin{array}{r} 9.14 \\ -\ 2.376 \\ \hline \end{array}$

1. 7.62 + 5.97 + 3.6 =

2. $\begin{array}{r} .9 \\ .7 \\ +\ .6 \\ \hline \end{array}$

3. 27 - .27 =

4. 4.9 - 1.666 =

5. 36.19 + 24 + 32.916 =

6. 4.963 - 2.9 =

7. .9 + 5 + .7 =

8. 55.828 + 6.97 + 3.42 =

9. 17.6 - 8.27 =

10. $\begin{array}{r} 47.2 \\ 3.647 \\ 5.23 \\ +\ 16.924 \\ \hline \end{array}$

1.
2.
3.
4.
5.
6.
7.
8.
9.
10.

Score

Problem Solving

The Jenkins family drove 1250.5 miles in three days. The first day they drove 550.25 miles and on the second day they drove 352.3 miles. How far did they drive on the third day?

Review Exercises

1. 34×5

2. 47×33

3. 307×36

4. $7.63 + 3.4 + 9.66$

5. $7.2 - 2.637$

6. $7.5 + 3 + 2.6 + 3.4 =$

Helpful Hints

Multiply as you would with whole numbers. Find the number of decimal places and place the decimal point properly in the product.

Examples:

2.32 ← 2 places	7.6 ← 1 place
x 6	x 23
13.92 ← 2 places	228
	1520
	174.8 ← 1 place

S1. 2.46×3

S2. 2.3×16

1. $.643 \times 3$

2. 3.66×4

3. $.16 \times 43$

4. 2.36×24

5. 1.4×16

6. 3.45×16

7. 7.63×43

8. 1.432×7

9. $.41 \times 73$

10. $.046 \times 27$

1.
2.
3.
4.
5.
6.
7.
8.
9.
10.
Score

Problem Solving

A farmer can harvest 7.6 tons of crops per day.
How many tons of crops can be harvested in 5 days?

Review Exercises

1. 4.26×3

2. 2.3×14

3. 7.213×5

4. $.032 \times 23$

5. $7.1 - 2.367$

6. $7.6 + .72 + 5.1 + 6.327$

Helpful Hints

Use what you have learned to solve the following problems.

* Remember to write the decimal in the answer in the proper location.

S1. 3.47×5

S2. 4.4×23

1. $.723 \times 4$

2. 3.72×6

3. $.27 \times 46$

4. 3.62×24

5. 4.2×16

6. 3.45×23

7. 7.124×8

8. 2.97×6

9. 7.5×76

10. $.137 \times 24$

1.
2.
3.
4.
5.
6.
7.
8.
9.
10.
Score

Problem Solving

Shirts are on sale for $20.15 each.
What would be the cost of five shirts?

Review Exercises

1. $7.25 + 3.7 + 4.637$

2. $7.132 - 1.476$

3. 1.15×5

4. $.24 \times 16$

5. 6 - .37 =

6. 7.21 + 9 + 6.426 =

Helpful Hints

Multiply as you would with whole numbers. Find the number of decimal places and place the decimal point properly in the product.

Examples:

2.63 ← 2 places
x .3 ← 1 place
.789 ← 3 places

.724 ← 3 places
x .23 ← 2 places
2172
14480
.16652 ← 5 places

S1. $3.6 \times .7$

S2. 3.24×2.4

1. 3.6×3.2

2. $2.09 \times .22$

3. $.642 \times .33$

4. $.23 \times 3.8$

5. $2.03 \times .07$

6. $.422 \times 23.2$

7. $.003 \times 0.8$

8. 5.6×3.4

9. 63.5×2.35

10. $12.3 \times .006$

1.
2.
3.
4.
5.
6.
7.
8.
9.
10.
Score

Problem Solving

One pound of shrimp costs $3.20.
How much will 2.5 pounds cost?

Review Exercises

1. $\begin{array}{r} 6.42 \\ \times \quad 6 \\ \hline \end{array}$

2. Write 3.0026 in words.

3. Write 3.007 as a mixed numeral.

4. Write two and eleven thousandths as a decimal.

5. Write $\frac{16}{100,000}$ in words.

6. $\begin{array}{r} 4.6 \\ \times \ .23 \\ \hline \end{array}$

Helpful Hints

Use what you have learned to solve the following problems.

* Remember to place the decimal properly in the answer.

S1. $\begin{array}{r} .324 \\ \times \ .7 \\ \hline \end{array}$

S2. $\begin{array}{r} 3.26 \\ \times \ 4.2 \\ \hline \end{array}$

1. $\begin{array}{r} 5.3 \\ \times \ 4.6 \\ \hline \end{array}$

2. $\begin{array}{r} 2.09 \\ \times \ .33 \\ \hline \end{array}$

3. $\begin{array}{r} .426 \\ \times \ .44 \\ \hline \end{array}$

4. $\begin{array}{r} 4.07 \\ \times \ .06 \\ \hline \end{array}$

5. $\begin{array}{r} .27 \\ \times \ 4.6 \\ \hline \end{array}$

6. $\begin{array}{r} .433 \\ \times \ 2.73 \\ \hline \end{array}$

7. $\begin{array}{r} .007 \\ \times \ 0.6 \\ \hline \end{array}$

8. $\begin{array}{r} 6.4 \\ \times \ .43 \\ \hline \end{array}$

9. $\begin{array}{r} 6.51 \\ \times \ 2.34 \\ \hline \end{array}$

10. $\begin{array}{r} 12.7 \\ \times \ .008 \\ \hline \end{array}$

1.
2.
3.
4.
5.
6.
7.
8.
9.
10.
Score

Problem Solving

Five friends are going to attend a concert together.
If tickets are $17.50 each, what is the total cost of tickets?

Review Exercises

1. 2.14 x 3 =

2. 2.13 x 1.7 =

3. .002 x .003 =

4. 3.63 + 7.75 + 4.62 =

5. 5.1 - 3.243 =

6. $3.54 x 5 =

Helpful Hints

To multiply by 10, move the decimal point one place to the right; by 100, two places to the right; and by 1,000, three places to the right

Examples:

10 x 3.36 = 33.6
100 x 3.36 = 336
1000 x 3.36 = 3360*

*Sometimes placeholders are necessary.

S1. 10 x 3.2 =

S2. 1,000 x 7.39 =

1. 100 x .936 =

2. 1,000 x 72.6 =

3. 100 x 1.6 =

4. 7.362 x 100 =

5. 7.28 x 1000 =

6. 100 x .7 =

7. 100 x .376 =

8. 1000 x .39 =

9. 100 x .733 =

10. 10 x 7.63 =

1.
2.
3.
4.
5.
6.
7.
8.
9.
10.

Score

Problem Solving

If tickets to a concert are $18.50, how much would 1,000 tickets cost?

Review Exercises

1. Write 2.763 in words.

2. Write seven hundred-thousandths as a decimal

3. 5 - .55 =

4. Write $\frac{16}{100,000}$ as a decimal.

5. .723 x .5

6. 7.6 + 44.27 + 1.93 =

Helpful Hints

Use what you have learned to solve the following problems.
* Sometimes place holders are necessary

S1. 100 x 3.7 =

S2. 1,000 x 5.36 =

1. 100 x 3.27 =

2. 1,000 x 56.7 =

3. 100 x 1.9 =

4. 7.364 x 100 =

5. .976 x 1000 =

6. 1,000 x .75 =

7. 10 x 7.3 =

8. 1000 x .387

9. 100 x 83.3 =

10. 8.42 x 10 =

1.
2.
3.
4.
5.
6.
7.
8.
9.
10.
Score

Problem Solving

If individual floor tiles weigh 2.5 pounds each, what is the weight of 1,000 floor tiles?

Review Exercises

1. 23.2×6

2. $.36 \times 25$

3. 32.5×100

4. $7.06 \times .6$

5. $.003 \times .003$

6. 4.23×5.1

Helpful Hints

Use what you have learned to solve the following problems.
* Be careful when placing the decimal point in the product.
* Sometimes place holders are necessary.

S1. $.342 \times 7$

S2. $42.3 \times .36$

1. $.23 \times 14$

2. $.29 \times 1.6$

3. $1.34 \times .362$

4. 10 x 2.6 =

5. 2.63×1.2

6. 100 x 26.3 =

7. $.003 \times 3.6$

8. $.65 \times 5.5$

9. 1.67×33

10. $.67 \times .063$

1.
2.
3.
4.
5.
6.
7.
8.
9.
10.
Score

Problem Solving

If Omar earns $16 per hour, how much will he earn in 4.5 hours?

Review Exercises

1. 2.76 + 5 + 3.99 =

2. 7.1 - 2.334 =

3. 62.83 + 7.9 + 4.652

4. 15 - 2.76 =

5. 15 - .12 =

6. 2.365 x 1000

Helpful Hints

Use what you have learned to solve the following problems.
* Be careful when placing the decimal point in the product.
* Practice reading the answers to yourself.

S1. 4.26 x 6

S2. .062 x 5.8

1. 42 x 2.3

2. .47 x 6.3

3. 2.09 x .003

4. 100 x 3.6 =

5. 37.2 x 2.4

6. .005 x .001

7. 26.4 x 5.23

8. 42.9 x 1,000 =

9. 2.17 x 16

10. .046 x 0.63

1.
2.
3.
4.
5.
6.
7.
8.
9.
10.
Score

Problem Solving

A man bought three bags of chips at $.59 each and a pizza for $11.75, how much did he spend in all?

Review Exercises

1. $6\overline{)134}$
2. $5\overline{)1515}$
3. $22\overline{)2442}$
4. $22\overline{)7963}$
5. $18\overline{)2376}$
6. $25\overline{)5075}$

Helpful Hints

Divide as you would with whole numbers. Place the decimal point directly up.

Examples:

$$\begin{array}{r} 2.8 \\ 3\overline{)8.4} \\ -\underline{6\downarrow} \\ 24 \\ -\underline{24} \\ 0 \end{array} \qquad \begin{array}{r} .084 \\ 3\overline{).252} \\ -\underline{24\downarrow} \\ 12 \\ -\underline{12} \\ 0 \end{array}$$

S1. $3\overline{)1.32}$

S2. $8\overline{)14.4}$

1. $3\overline{)59.1}$
2. $7\overline{)22.47}$
3. $34\overline{)19.38}$
4. $70.3 \div 19 =$
5. $4\overline{)24.32}$
6. $6\overline{)245.4}$
7. $26\overline{)8.424}$
8. $16\overline{)2.56}$
9. $12.72 \div 6 =$
10. $21\overline{)42.84}$

Answers
1.
2.
3.
4.
5.
6.
7.
8.
9.
10.
Score

Problem Solving

If five loaves of bread cost $9.75, how much does one loaf cost?

Review Exercises

1. $3\overline{)13.2}$
2. $5\overline{)1.725}$
3. $22\overline{)35.86}$
4. $7.12 - 1.637$
5. $.36 \times 2.4$
6. $36.8 + 4.92 + 36.57$

Helpful Hints

Use what you have learned to solve the following problems.
* Place the decimal point directly up.

S1. $4\overline{).928}$

S2. $7\overline{)18.76}$

1. $3\overline{)1.53}$
2. $8\overline{)32.48}$
3. $32\overline{)17.92}$
4. $41\overline{)229.6}$
5. $3\overline{)6.912}$
6. $65\overline{)14.95}$
7. $24\overline{)15.12}$
8. $32\overline{)5.12}$
9. $6\overline{).738}$
10. $21\overline{)85.68}$

1.
2.
3.
4.
5.
6.
7.
8.
9.
10.
Score

Problem Solving

Al's times for the one hundred yard dash were 11.8 seconds, 12.5 seconds, and 12.3 seconds. What was his average time?

Review Exercises

1. $3\overline{)22.14}$
2. $15\overline{)4.515}$
3. 100 x 2.7 =
4. 75 - 36.2 =
5. Find the sum of 256.7 and 22.96.
6. Write 2.1762 as a mixed numeral.

Helpful Hints

Sometimes placeholders are necessary when dividing decimals.

Examples:

$$\begin{array}{r} .05 \\ 3\overline{).15} \\ -\underline{15} \\ 0 \end{array} \qquad \begin{array}{r} .003 \\ 15\overline{).045} \\ -\underline{45} \\ 0 \end{array}$$

S1. $5\overline{).0135}$

S2. $13\overline{).247}$

1. $7\overline{).0049}$
2. $3\overline{).036}$
3. $4\overline{).224}$
4. $13\overline{).468}$
5. $22\overline{).946}$
6. $9\overline{).567}$
7. $52\overline{)1.196}$
8. $18\overline{).396}$
9. $9\overline{).027}$
10. $12\overline{).816}$

Answers
1.
2.
3.
4.
5.
6.
7.
8.
9.
10.
Score

Problem Solving

A man bought two hammers for $5.25 each and three saws for $7.50 each. What was the total cost?

Review Exercises

1. Write 2.003 in words.

2. 32.6 + 5. + 72.613

3. 76.2 - 9.213

4. 62.3 x 7

5. $4\overline{)53.2}$

6. $22\overline{)3.784}$

Helpful Hints

Use what you have learned to solve the following problems.
* Be careful to include placeholders when necessary.

S1. $6\overline{).330}$

S2. $13\overline{).494}$

1. $9\overline{).0036}$

2. $7\overline{).014}$

3. $4\overline{).288}$

4. $12\overline{).552}$

5. $42\overline{)1.428}$

6. $6\overline{).324}$

7. $62\overline{)3.534}$

8. $18\overline{).792}$

9. $8\overline{).0064}$

10. $12\overline{).408}$

1.
2.
3.
4.
5.
6.
7.
8.
9.
10.
Score

Problem Solving

Shirts are on sale for $9.50 each. If the regular price is $11.25, how much would be saved by buying three shirts on sale?

Review Exercises

1. 22.6×3

2. $.342 \times .06$

3. $.36 \times 1{,}000 =$

4. $5\overline{).015}$

5. $12\overline{).132}$

6. Write six and fifteen hundred-thousandths as a decimal.

Helpful Hints

Sometimes zeroes need to be added to the dividend to complete the problem.

* Sometimes it is necessary to add more than one zero.

Examples: $5\overline{)1.3}$ $15\overline{)2.7}$

$$\begin{array}{r} .26 \\ 5\overline{)1.30} \\ -1\,0\downarrow \\ 30 \\ -\,30 \\ 0 \end{array} \qquad \begin{array}{r} .18 \\ 15\overline{)2.70} \\ -1\,5\downarrow \\ 120 \\ -\,120 \\ 0 \end{array}$$

S1. $5\overline{)1.7}$

S2. $25\overline{)1.5}$

1. $2\overline{).13}$

2. $5\overline{)3.1}$

3. $22\overline{)45.1}$

4. $24\overline{)3.6}$

5. $5\overline{)0.2}$

6. $95\overline{)3.8}$

7. $20\overline{)2.4}$

8. $4\overline{)6.3}$

9. $5\overline{).03}$

10. $5\overline{)2.09}$

1.
2.
3.
4.
5.
6.
7.
8.
9.
10.
Score

Problem Solving

A woman bought three gallons of gas at $3.29 per gallon. If she paid with a $20.00 bill, what would be the change?

Review Exercises

1. $5\overline{)\$7.95}$

2. $\$7.90 \times 8$

3. $2\overline{).13}$

4. 42.6 + 3.92 + .96 =

5. 55.23 - 36.712 =

6. $95\overline{)7.6}$

Helpful Hints

Use what you have learned to solve the following problems.
* Use placeholders when necessary.
* Add zeroes to the dividend when necessary.

S1. $5\overline{).37}$

S2. $22\overline{)49.5}$

1. $2\overline{).17}$

2. $5\overline{).39}$

3. $20\overline{)1.2}$

4. $5\overline{).8}$

5. $5\overline{)3.19}$

6. $18\overline{)2.7}$

7. $4\overline{)5.8}$

8. $2\overline{).37}$

9. $4\overline{)5.4}$

10. $15\overline{).6}$

1.
2.
3.
4.
5.
6.
7.
8.
9.
10.
Score

Problem Solving

If 3 pounds of butter is $5.43, what is the price per pound?

Review Exercises

1. $2\overline{)\,.19}$

2. $5\overline{)\,1.7}$

3. $6\overline{)\,.0012}$

4. Change $5\frac{13}{10{,}000}$ to a decimal.

5. Write .00135 as a fraction.

6. $\begin{array}{r} .7 \\ .9 \\ +\ .6 \\ \hline \end{array}$

Helpful Hints

When dividing a decimal by another decimal, move the decimal point in the divisor the number of places necessary to make it a whole number. Move the decimal point in the dividend the same number of places.

Examples:

$$\begin{array}{r} .8 \\ .3\overline{)\,.2.4} \\ -\,2\,4 \\ \hline 0 \end{array} \qquad \begin{array}{r} 9\,50.^{*} \\ .03.\overline{)\,28.50.} \\ -\,27 \\ \hline 1\,5 \\ -1\,5 \\ \hline 0 \end{array}$$

*Sometimes placeholders are necessary

S1. $.7\overline{)\,2.73}$

S2. $.15\overline{)\,.036}$

1. $.3\overline{)\,2.4}$

2. $.03\overline{)\,5.1}$

3. $.9\overline{)\,.378}$

4. $.04\overline{)\,3.2}$

5. $.06\overline{)\,.324}$

6. $2.1\overline{)\,6.72}$

7. $.26\overline{)\,.962}$

8. $.18\overline{)\,.576}$

9. $.04\overline{)\,2.3}$

10. $.12\overline{)\,1.104}$

1.
2.
3.
4.
5.
6.
7.
8.
9.
10.
Score

Problem Solving

Five cans of beans cost $2.75. One can of tuna costs $.79. How much would it cost for one can of beans and two cans of tuna?

Review Exercises

1. $0.23 + 1.47 + 0.37 =$
2. $8.7 - 2.79 =$
3. $1.2 \times 3.42 =$
4. $1.3 \div 2 =$
5. $.17 \div 5 =$
6. $12\overline{)5.16}$

Helpful Hints

Use what you have learned to solve the following problems.
* Use placeholders when necessary.
* Add zeroes to the dividend when necessary.

S1. $.8\overline{)2.5}$

S2. $.15\overline{)76.2}$

1. $.2\overline{).25}$
2. $.8\overline{).024}$
3. $.15\overline{).762}$
4. $.04\overline{)9.2}$
5. $.02\overline{).4}$
6. $.05\overline{).325}$
7. $2.1\overline{).693}$
8. $.15\overline{)7.5}$
9. $.53\overline{)4.876}$
10. $.03\overline{).072}$

1.
2.
3.
4.
5.
6.
7.
8.
9.
10.
Score

Problem Solving

A stack of tiles is 27.5 inches tall. If each tile is .5 inches tall, how many tiles are in the stack?

Review Exercises

1. $3\overline{)4.8}$
2. $.04\overline{).6}$
3. $.5\overline{).17}$
4. $.12\overline{)36}$
5. $.29\overline{)2.929}$
6. $1.3\overline{)6.76}$

Helpful Hints

To change fractions to decimals, divide the numerator by the denominator. Add as many zeroes as necessary.

Examples:

$\frac{3}{4}$

$$\begin{array}{r} .75 \\ 4\overline{)3.00} \\ -2\,8\downarrow \\ \hline -20 \\ -20 \\ \hline 0 \end{array}$$

$\frac{3}{8}$

$$\begin{array}{r} .375 \\ 8\overline{)3.000} \\ -2\,4\downarrow\downarrow \\ \hline 60 \\ -56 \\ \hline 40 \\ -40 \\ \hline 0 \end{array}$$

Change each fraction to a decimal.

S1. $\frac{3}{4}$

S2. $\frac{5}{8}$

1. $\frac{3}{5}$
2. $\frac{1}{4}$
3. $\frac{2}{5}$
4. $\frac{7}{8}$
5. $\frac{11}{20}$
6. $\frac{13}{25}$
7. $\frac{5}{8}$
8. $\frac{4}{20}$
9. $\frac{1}{5}$
10. $\frac{7}{10}$

1.
2.
3.
4.
5.
6.
7.
8.
9.
10.
Score

Problem Solving

A worker earned $700 and put .6 of it into his savings account. How much did he put into his savings account?
(Hint: What operation does "of" usually mean?)

Review Exercises

1. Which is the larger decimal, 2.3 or 2.199?

2. $\begin{array}{r} 7.73 \\ 14.2 \\ +\ 7.16 \\ \hline \end{array}$

3. $\begin{array}{r} .207 \\ \times\ 1.4 \\ \hline \end{array}$

4. $5 - .123 =$

5. $.5\overline{)2}$

6. $.03\overline{)1.5}$

Helpful Hints

Use what you have learned to solve the following problems.
* Add as many zeroes as necessary.

Change each fraction to a decimal.

S1. $\frac{3}{8}$

S2. $\frac{3}{4}$

1. $\frac{4}{5}$

2. $\frac{3}{16}$

3. $\frac{11}{25}$

4. $\frac{3}{20}$

5. $\frac{6}{8}$

6. $\frac{1}{5}$

7. $\frac{13}{50}$

8. $\frac{7}{10}$

9. $\frac{7}{25}$

10. $\frac{7}{16}$

1.
2.
3.
4.
5.
6.
7.
8.
9.
10.
Score

Problem Solving

Cans of peas are on sale for two for $.69.
How much would eight cans cost?

Review Exercises

1. $3.35 + 5 + 2.13 =$
2. $12 - 2.16 =$
3. $100 \times 1.236 =$
4. $7 \times 3.16 =$
5. $2.2 \times .006 =$
6. $5\overline{)1.9}$

Helpful Hints

Use what you have learned to solve the following problems.

* Add as many zeroes as necessary.
* Placeholders may be necessary.
* Place decimal points properly.

S1. $7\overline{).035}$

S2. $.06\overline{)2.4}$

1. $3\overline{)2.28}$

2. $5\overline{).37}$

3. $1.6\overline{).04}$

4. $.3\overline{)1.35}$

5. $.5\overline{).12}$

6. $.005\overline{)1.42}$

7. $.04\overline{)1.324}$

8. $2.1\overline{)34.02}$

9. Change $\frac{2}{5}$ to a decimal.

10. Change $\frac{5}{8}$ to a decimal.

1.
2.
3.
4.
5.
6.
7.
8.
9.
10.
Score

Problem Solving

Steak costs $ 3.80 per pound. How much does 2.5 pounds cost?

Review Exercises

1. $3\overline{)59.1}$
2. $16\overline{)2.56}$
3. $5\overline{).015}$
4. $4\overline{)6.3}$
5. $.06\overline{)3.24}$
6. $.21\overline{).672}$

Helpful Hints

Use what you have learned to solve the following problems.
* If necessary, refer to the "Helpful Hints" section from previous pages.

S1. $2\overline{)1.25}$

S2. $.07\overline{)4.9}$

1. $5\overline{).039}$
2. $3\overline{)2.34}$
3. $.02\overline{).146}$
4. $.24\overline{)1.2}$
5. $.3\overline{)6.375}$
6. $.005\overline{)1.43}$
7. $4.5\overline{)1.035}$
8. $.84\overline{).546}$
9. Change $\frac{3}{4}$ to a decimal.
10. Change $\frac{1}{4}$ to a decimal.

1.
2.
3.
4.
5.
6.
7.
8.
9.
10.
Score

Problem Solving

A man bought a car. He made a down payment of $1,200.00 and paid $300 per month for 36 months.
How much did he pay altogether?

Final Review of All Decimal Operations

1. $\begin{array}{r} 4.56 \\ 7.8 \\ + 3.976 \\ \hline \end{array}$

2. $.3 + 4.67 + 8.9 =$

3. $16.8 + 5 + 12.7 =$

4. $\begin{array}{r} 47.6 \\ - 19.7 \\ \hline \end{array}$

5. $\begin{array}{r} 9.2 \\ - 3.652 \\ \hline \end{array}$

6. $72 - 1.49 =$

7. $\begin{array}{r} 4.76 \\ \times 4 \\ \hline \end{array}$

8. $\begin{array}{r} 5.6 \\ \times 23 \\ \hline \end{array}$

9. $\begin{array}{r} .49 \\ \times 2.6 \\ \hline \end{array}$

10. $\begin{array}{r} .503 \\ \times 3.46 \\ \hline \end{array}$

11. $1,000 \times 3.19 =$

12. $100 \times 3.2 =$

13. $5 \overline{\smash{)}.79}$

14. $2 \overline{\smash{)}3.96}$

15. $.004 \overline{\smash{)}1.2}$

16. $.7 \overline{\smash{)}.224}$

17. $.15 \overline{\smash{)}.0045}$

18. $6.8 \overline{\smash{)}16.32}$

19. Change $\frac{5}{8}$ to a decimal.

20. Change $\frac{9}{16}$ to a decimal.

1.
2.
3.
4.
5.
6.
7.
8.
9.
10.
11.
12.
13.
14.
15.
16.
17.
18.
19.
20.
Score

Final Test of all Decimal Operations

1.
$$\begin{array}{r} 5.62 \\ 15.7 \\ 8.236 \\ +\ 12.16 \\ \hline \end{array}$$

2. $7.6 + 5 + .9 + 2.72 =$

3. $.7 + .6 + .9 + .7 =$

4. $72 - .72 =$

5.
$$\begin{array}{r} 72.6 \\ -\ 19.723 \\ \hline \end{array}$$

6. $.3 - .216 =$

7.
$$\begin{array}{r} .937 \\ \times\ \ 5 \\ \hline \end{array}$$

8.
$$\begin{array}{r} 15 \\ \times\ 5.6 \\ \hline \end{array}$$

9.
$$\begin{array}{r} .72 \\ \times\ 4.9 \\ \hline \end{array}$$

10.
$$\begin{array}{r} .207 \\ \times\ .69 \\ \hline \end{array}$$

11. $4.1 \times 100 =$

12. $3.762 \times 1{,}000 =$

13. $5\overline{)3.95}$

14. $5\overline{).032}$

15. $.9\overline{).225}$

16. $.005\overline{)30}$

17. $.25\overline{)117.5}$

18. $.12\overline{).2544}$

19. Change $\frac{1}{8}$ to a decimal.

20. Change $\frac{7}{8}$ to a decimal.

1.
2.
3.
4.
5.
6.
7.
8.
9.
10.
11.
12.
13.
14.
15.
16.
17.
18.
19.
20.
Score

Review Exercises

1. Change $\frac{17}{100}$ to a decimal.
2. Change $\frac{9}{10}$ to a decimal.
3. Write .07 as a fraction.
4. Write .7 as a fraction.
5. Write $6\frac{7}{100}$ in words.
6. Write 10.9 in words.

Helpful Hints

Percent means "per hundred" or "hundredths." If a fraction is expressed as hundredths, it can easily be written as a percent.

Examples:

$\frac{7}{100} = 7\%$　$\frac{3}{10} = 30\%$　$\frac{19}{100} = 19\%$

Change each of the following to a percent.

S1. $\frac{17}{100} =$　S2. $\frac{9}{10} =$　1. $\frac{6}{100} =$

2. $\frac{99}{100} =$　3. $\frac{3}{10} =$　4. $\frac{64}{100} =$

5. $\frac{67}{100} =$　6. $\frac{1}{100} =$　7. $\frac{7}{10} =$

8. $\frac{14}{100} =$　9. $\frac{80}{100} =$　10. $\frac{62}{100} =$

1.
2.
3.
4.
5.
6.
7.
8.
9.
10.
Score

Problem Solving

Ron is taking a trip of 252 miles. If his car gets 21 miles per gallon of gas, how many gallons of gas will the car consume? If gas costs $3.25 per gallon, how much will Ron spend on gas for the entire trip?

Review Exercises

1. $7\overline{)\,.119}$

2. $.003\overline{)\,1.5}$

3. 100 x 3.4 =

4. 6.12 x .7 =

5. 7.63 + 52 + 9.64 =

6. 7.1 - 2.964 =

Helpful Hints

Use what you have learned to solve the following problems.

Change each of the following to a percent.

S1. $\frac{7}{10} =$

S2. $\frac{3}{100} =$

1. $\frac{19}{100} =$

2. $\frac{87}{100} =$

3. $\frac{6}{10} =$

4. $\frac{63}{100} =$

5. $\frac{19}{100} =$

6. $\frac{2}{100} =$

7. $\frac{48}{100} =$

8. $\frac{14}{100} =$

9. $\frac{5}{10} =$

10. $\frac{98}{100} =$

1.
2.
3.
4.
5.
6.
7.
8.
9.
10.
Score

Problem Solving

Ellen worked 25 hours and earned 8 dollars per hour. If she wants to buy a bike for $372, how much more money does she need?

Review Exercises

1. $\frac{7}{10}$ = ________%

2. $\frac{72}{100}$ = ________%

3. $.03\overline{)12}$

4. Change $\frac{5}{8}$ to a decimal.

5. Change $\frac{3}{5}$ to a decimal.

6. $.003 \times .002$

Helpful Hints

"Hundredths" = Percent

Decimals can easily be changed to percents.

Examples:

.27 = 27% .9 = .90 = 90% .124 = 12.4%

* Move the decimal point twice to the right and add a percent symbol.

Change each of the following to a percent.

S1. .37

S2. .7

1. .93

2. .02

3. .2

4. .09

5. .6

6. .665

7. .89

8. .6

9. .334

10. .8

1.
2.
3.
4.
5.
6.
7.
8.
9.
10.
Score

Problem Solving

There are 16 fluid ounces in a pint.
How many fluid ounces are there in .6 pints?

Review Exercises

1. Write 3.0196 as a mixed numeral.

2. Write .0021 in words.

3. Write $7\frac{7}{10,000}$ as a decimal.

4. $.15\overline{).4545}$

5. $.5\overline{)2}$

6. $1.6\overline{).352}$

Helpful Hints

Use what you have learned to solve the following problems.
* Move the decimal twice to the right and add a percent symbol.

Change each of the following to a percent.

S1. .09

S2. .348

1. .90

2. .09

3. .7

4. .097

5. .6

6. .007

7. .87

8. .3

9. .445

10. .4

1.
2.
3.
4.
5.
6.
7.
8.
9.
10.
Score

Problem Solving

A school has 480 students. If .25 of them ride the bus to school, how many students take the bus? (Hint: What does "of" mean?)

Review Exercises

1. $7.9 - .79 =$

2. $5 - 2.78 =$

3. $3.46 + 15 + .78 =$

4. $.9 + .76 + .73 + .8 =$

5. $5.13 - 2.667$

6. $17.54 + 6.723 + 36.124$

Helpful Hints

Percents can be expressed as decimals and fractions. The fraction form may sometimes be reduced to its lowest terms.

Examples: $25\% = .25 = \frac{25}{100} = \frac{1}{4}$

$8\% = .08 = \frac{8}{100} = \frac{2}{25}$

Change each percent to a decimal and to a fraction reduced to its lowest terms.

S1. 20% = . = ___

S2. 9% = . = ___

1. 16% = . = ___

2. 6% = . = ___

3. 75% = . = ___

4. 40% = . = ___

5. 1% = . = ___

6. 45% = . = ___

7. 12% = . = ___

8. 5% = . = ___

9. 50% = . = ___

10. 13% = . = ___

1.
2.
3.
4.
5.
6.
7.
8.
9.
10.
Score

Problem Solving

If 25% of the students at Eaton School take the bus, what fraction of the students take the bus? (Reduce your answer to its lowest terms.)

Review Exercises

1. $\begin{array}{r} 12.7 \\ \times \quad 5 \\ \hline \end{array}$

2. $\begin{array}{r} 24.5 \\ \times \ .75 \\ \hline \end{array}$

3. $\begin{array}{r} .008 \\ \times \ .03 \\ \hline \end{array}$

4. $2\overline{).15}$

5. $1.2\overline{).2424}$

6. $.18\overline{).468}$

Helpful Hints

Use what you have learned to solve the following problems.
* Be sure fractions are reduced to lowest terms.

Change each percent to a decimal and to a fraction reduced to its lowest terms.

S1. 50% = . = ___

S2. 5% = . = ___

1. 8% = . = ___

2. 80% = . = ___

3. 24% = . = ___

4. 11% = . = ___

5. 2% = . = ___

6. 70% = . = ___

7. 9% = . = ___

8. 90% = . = ___

9. 17% = . = ___

10. 14% = . = ___

1.
2.
3.
4.
5.
6.
7.
8.
9.
10.
Score

Problem Solving

A yard is in the shape of a rectangle that is 10 feet wide and 12 feet long. To build a fence it costs $20 per foot. How much would it cost to build a fence around the yard? (Hint: Make a sketch.)

Review Exercises

1. Change 80% to a decimal.

2. Change 7% to a decimal.

3. Change 25% to fraction reduced to lowest terms.

4. $156 \times .7$

5. $400 \times .32$

6. $300 \times .06$

Helpful Hints

To find the percent of a number you may use either fractions or decimals. Use what is the most convenient.

Examples: Find 25% of 60
.25 x 60

$$\begin{array}{r} 60 \\ \times .25 \\ \hline 300 \\ 120 \\ \hline 15.00 \end{array}$$

OR

$$\frac{25}{100} = \frac{1}{4}$$

$$\frac{1}{\cancel{4}_1} \times \frac{\cancel{60}^{15}}{1} = \frac{15}{1} = 15$$

S1. Find 70% of 25.

S2. Find 50% of 300.

1. Find 6% of 72.

2. Find 60% of 85.

3. Find 25% of 60.

4. Find 45% of 250.

5. Find 10% of 320.

6. Find 40% of 200.

7. Find 4% of 250.

8. Find 90% of 240.

9. Find 75% of 150.

10. Find 2% of 660.

1.
2.
3.
4.
5.
6.
7.
8.
9.
10.
Score

Problem Solving

A train traveled 400 miles in 2.5 hours.
What was its average speed per hour?

Review Exercises

1. $.12\overline{).048}$

2. $5\overline{)2}$

3. $.2\overline{).13}$

4. Find .9 of 45

5. Write $\frac{3}{8}$ as a decimal.

6. Write 3.0016 in words.

Helpful Hints

Use what you have learned to solve the following problems.
* Use fractions or decimals, depending on which is the most convenient.

S1. Find 6% of 400.

S2. Find 60% of 400.

1. Find 4% of 80.

2. Find 70% of 550.

3. Find 7% of 550.

4. Find 25% of 200.

5. Find 20% of 250.

6. Find 60% of 310.

7. Find 5% of 220.

8. Find 80% of 300.

9. Find 75% of 160.

10. Find 25% of 24.

1.
2.
3.
4.
5.
6.
7.
8.
9.
10.
Score

Problem Solving

Rosa's test scores were 80, 90, 86, and 80.
What was her average test score?

Review Exercises

1. Find 12% of 60.

2. Find 90% of 320.

3. Find 25% of 40.

4. Change $\frac{3}{4}$ to a decimal.

5. Change 50% to a fraction reduced to the lowest terms.

6. Find .3 of 75.

Helpful Hints

When finding the percent of a number in a word problem, you can change the percent to a fraction or a decimal. Always express your answer in a short phrase or sentence.

Example:
A team played 60 games and won 75% of them. How many games did they win?

Find 75% of 60.
.75 x 60

$$\begin{array}{r} 60 \\ \underline{\times .75} \\ 300 \\ \underline{420} \\ 45.00 \end{array}$$

OR

$$\frac{75}{100} = \frac{3}{4}$$

$$\frac{3}{\cancel{4}_1} \times \frac{\cancel{60}^{15}}{1} = \frac{45}{1} = 45$$

Answer: The team won 45 games.

S1. George took a test with 40 problems. If he got 15% of the problems correct, how many problems did he get correct?

S2. If 6% of the 500 students enrolled in a school are absent, then how many students are absent?

1. A worker earned 120 dollars and put 70% of it into the bank. How many dollars did he put into the bank?

2. A car costs $9,000. If Mr. Smith has saved 30% of this amount, how much did he save?

3. Steve took a test with 60 problems. If he got 70% of the problems correct how many of the problems did he get incorrect?

4. A family's monthly income is $3,000. If 20% of this amount is spent on food, how much money is spent on food?

5. There are 30 students in a class. If 60% of the class is boys, how many girls are in the class?

6. A house that costs $200,000 requires a 20% down payment. How many dollars are required for the down payment?

7. If a car costs $8,000 and loses 30% of its value in one year, how much will the car be worth in one year?

8. A coat is priced $50. If the sales tax is 7% of the price, how much is the sales tax? What is the total cost including sales tax?

9. 25% of the 600 students at Madison School take instrumental music. How many students are taking instrumental music?

10. A family spends 20% of its income for food and 30% for housing. If its monthly income is $3,000, how much is spent each month on food and housing?

1.
2.
3.
4.
5.
6.
7.
8.
9.
10.
Score

Problem Solving

A man had $362.00 in the bank. On Monday he withdrew $92.00, on Tuesday he deposited $76.00, and on Wednesday he withdrew $49.00. How much does he now have in the bank?

Review Exercises

1. Find 20% of 240.

2. Find 2% of 240.

3. Find $\frac{3}{4}$ of 240.

4. Change $\frac{1}{5}$ to a decimal.

5. $$\begin{array}{r} .9 \\ .6 \\ +\ .8 \\ \hline \end{array}$$

6. 7 - 5.55 =

Helpful Hints

Use what you have learned to solve the following problems.
* Use fractions or decimals depending on which is most convenient.
* Express your answer in a short phrase or sentence.

S1. 40 people in a class take a test. If 80% passed the test, how many passed?

S2. In a class of 30 people 40% are boys. How many are girls?

1. A bakery made 600 cookies and sold 90% of them. How many cookies were sold?

2. A bag has a mixture of 120 white and red marbles. If 40% of the marbles are red, how many white marbles are there?

3. In a school of 800 students, 60% ride the bus. How many ride the bus?

4. A book costs $12.00. If it was on sale for 30% off, how much would be saved buying it on sale?

5. Marco is buying a car priced at $12,000. If he needs a down payment of 30%, how much is the down payment?

6. Steve earns $600 and saves 20% of it. How much does he spend?

7. Allie's bill at a restaurant was $45.00. If she wanted to leave a 20% tip, how much should she leave?

8. Bill earns $800. If 40% goes to rent and 20% goes to his car payment, what is the total cost for his rent and car payment?

9. 150 math students took a test and 80% passed. How many did not pass?

10. A school has 600 student and 60% are boys. How many boys are there in the school? How many girls?

1.
2.
3.
4.
5.
6.
7.
8.
9.
10.
Score

Problem Solving

Two pounds of beef costs $2.50. How much does six pounds cost?

Review Exercises

1. $.05\overline{).245}$

2. Find 20% of 36.

3. Find 2% of 360.

4. 27.2 - 18.76 =

5. Change $7\frac{9}{1,000}$ to a decimal.

6. Change $\frac{9}{100}$ to a decimal.

Helpful Hints

To change a fraction to a percent, first change the fraction to a decimal, then change the decimal to a percent. Move the decimal twice to the right and add a percent symbol.

Examples: $\frac{3}{4}$ $4\overline{)3.00}$ = .75 = 75%; - 2.8, - 20, - 20, 0

$\frac{16}{20} = \frac{4}{5}$ $5\overline{)4.00}$ = .80 = 80%; - 4.0, 0

* Sometimes the fraction can be reduced further.

Change each of the following to a percent.

S1. $\frac{1}{5} =$

S2. $\frac{12}{15} =$

1. $\frac{3}{5} =$

2. $\frac{1}{2} =$

3. $\frac{1}{10} =$

4. $\frac{9}{12} =$

5. $\frac{15}{20} =$

6. $\frac{15}{25} =$

7. $\frac{1}{4} =$

8. $\frac{24}{30} =$

9. $\frac{18}{24} =$

10. $\frac{4}{20} =$

1.
2.
3.
4.
5.
6.
7.
8.
9.
10.
Score

Problem Solving

Three hundred sixty people work for a company. Forty percent of them carpool to work. Find how many people carpool to work.

Review Exercises

1. $\begin{array}{r} 33.3 \\ 4.44 \\ +\ 55.55 \\ \hline \end{array}$

2. 15 - .15 =

3. $\begin{array}{r} 2.17 \\ \times\ \ .7 \\ \hline \end{array}$

4. $5\overline{)4}$

5. $.07\overline{).777}$

6. $.14\overline{).0294}$

Helpful Hints

Use what you have learned to solve the following problems.
* Some fractions can be reduced further.

Change each of the following to a percent.

S1. $\frac{20}{25} =$

S2. $\frac{45}{60} =$

1. $\frac{4}{8} =$

2. $\frac{7}{10} =$

3. $\frac{6}{24} =$

4. $\frac{5}{25} =$

5. $\frac{5}{20} =$

6. $\frac{4}{16} =$

7. $\frac{5}{8} =$

8. $\frac{30}{60} =$

9. $\frac{9}{36} =$

10. $\frac{3}{8} =$

1.
2.
3.
4.
5.
6.
7.
8.
9.
10.
Score

Problem Solving

A worker completed $\frac{15}{20}$ of his project.
What percent of his project has been completed?

Review Exercises

1. Find 15% of 310.
2. Find 20% of 120.
3. $8\overline{)6}$
4. Change $\frac{1}{2}$ to a percent.
5. Find .9 of 150.
6. $.05\overline{)30}$

Helpful Hints

When finding the percent, first write a fraction, change the fraction to a decimal, then change the decimal to a percent. * "Is" means =.

Examples:

4 is what percent of 16?

$\frac{4}{16} = \frac{1}{4}$

$$\begin{array}{r} .25 = 25\% \\ 4\overline{)1.00} \\ -\ 8\downarrow \\ -\ 20 \\ -\ 20 \\ \hline 0 \end{array}$$

5 is what percent of 25?

$\frac{5}{25} = \frac{1}{5}$

$$\begin{array}{r} .20 = 20\% \\ 5\overline{)1.00} \\ -\ 1.0\downarrow \\ \hline 00 \end{array}$$

Change each of the following to a percent.

S1. 3 is what percent of 12?

S2. 15 is what percent of 20?

1. 7 is what percent of 28?
2. 20 is what percent of 25?
3. 40 = what percent of 80?
4. 18 is what percent of 20?
5. 12 is what percent of 20?
6. 9 is what percent of 12?
7. 15 = what percent of 20?
8. 24 is what percent of 32?
9. 400 is what percent of 500?
10. 19 is what percent of 20?

1.
2.
3.
4.
5.
6.
7.
8.
9.
10.
Score

Problem Solving

A rancher has 800 cows. If he sells 60% of them, how many will he have left?

Review Exercises

1. Write seven and twenty-eight hundredths as a decimal.

2. Write three thousandths as a fraction.

3. Find 40% of 60.

4. Which is the larger decimal? .796 or .9

5. $.15\overline{)15}$

6. $\begin{array}{r} 2.08 \\ \underline{\times\ 1.6} \end{array}$

Helpful Hints

Use what you have learned to solve the following problems.
1. Write a fraction.
2. Change the fraction to a decimal.
3. Change the decimal to a percent.

Change each of the following to a percent.

S1. 9 is what percent of 12?

S2. 5 is what percent of 25?

1. 2 is what percent of 5?

2. 27 is what percent of 36?

3. 9 is what percent of 10?

4. 9 is what percent of 20?

5. 8 is what percent of 32?

6. 30 is what percent of 40?

7. 60 is what percent of 80?

8. 12 is what percent of 20?

9. 45 is what percent of 50?

10. 13 is what percent of 25?

1.
2.
3.
4.
5.
6.
7.
8.
9.
10.
Score

Problem Solving

A student took a test with 40 questions. If he got 90% of them correct, how many problems did he get correct?

Review Exercises

1. Find 40% of 280.
2. Find 5% of 60.
3. 3 is what % of 5?
4. 15 is what % of 25?
5. Write .471 as a percent.
6. Write $72\frac{601}{100,000}$ as a decimal.

Helpful Hints

When finding the percent first write a fraction, next change the fraction to a decimal, then change the decimal to a percent.

Example:

A team played 20 games and won 15 of them. What percent of the games did they win?

15 is what % of 20?

$\frac{15}{20} = \frac{3}{4}$

$$\begin{array}{r} .75 \\ 4\overline{)3.00} \\ -2\,8\downarrow \\ 20 \\ -20 \\ \hline 0 \end{array}$$

.75 = 75%

They won 75% of the games.

S1. A test had 25 questions. If Sam got 15 questions correct, what percent did she get correct?

S2. In a class of 20 students, 12 are girls. What percent of the class is girls?

1. On a spelling test with 25 words, Susan got 20 correct. What percent of the words did she get correct?

2. A worker earned 600 dollars. If she put 150 dollars into a savings account, what percent of her earnings did she put into a savings account?

3. A team played 16 games and won 12 of them. What percent did they lose?

4. A quarterback threw 35 passes and 21 were caught. What percent of the passes were caught?

5. $\frac{18}{20}$ of a class was present at school. What percent of the class was present?

6. A class has an enrollment of 30 students. If 24 are present, what percentage absent?

7. A team won 12 games and lost 13 games. What percent of the games played did they win?

8. A school has 300 students. If 120 of them are sixth graders, what percent are sixth graders?

9. On a math test with 50 questions Jill got 49 of them correct. What percent did she get correct?

10. A pitcher threw 12 pitches. If 9 of them were strikes, what percent were strikes?

1.
2.
3.
4.
5.
6.
7.
8.
9.
10.
Score

Problem Solving

A t.v. set costs $420. The sales tax is 8% of the price. What is the total price with tax included?

Review Exercises

1. Change $\frac{11}{20}$ to a percent.
2. Write .7 as a percent.
3. Find 20% of 40.
4. Find 25% of 80.
5. $.12\overline{)60}$
6. $.003\overline{)1.5}$

Helpful Hints

Use what you have learned to solve the following problems.
* Put your answer in a short phrase or sentence.
* If necessary, refer to the example on the previous page.

S1. In a class of 40 students, 10 of them received A's. What percent did not receive A's?

S2. On a spelling test with 12 words, Sean misspelled 3. What percent did he misspell?

1. A team played 15 games and won 12. What percent did they win?

2. A team played 25 games and won 20 of them. What percent of the games did they lose?

3. A class has 30 boys and 20 girls. What percent of the class is boys?

4. A worker earned 500 dollars and deposited 300 dollars of it into a savings account. What percent of his earnings did he deposit?

5. $\frac{9}{15}$ of the students in a school take the bus. What percent of the students take the bus.

6. Santiago has 35 fish. If 14 of them are goldfish, what percent of them are goldfish?

7. A basketball player shot 12 free throws and made 9 of them. What percent of the free throws did he miss?

8. In a class of 50 students, 45 of them have a home computer. What percent of them have a home computer?

9. A quarterback threw 24 passes and 18 were complete. What percent of the passes were completed?

10. In a survey of 40 people it is found that 12 of them have a pet. What percent of those surveyed have a pet?

1.
2.
3.
4.
5.
6.
7.
8.
9.
10.
Score

Problem Solving

Sandy worked five straight days and earned $250.75. What were her average daily earnings?

Review Exercises

1. Find 4% of 80.
2. Find 40% of 80.
3. 12 is what percent of 16?
4. 45 is what percent of 50?
5. 52 - 1.96 =
6. $.06\overline{)12}$

Helpful Hints

To find the whole when the part and the percent are known, simply change the equal sign " = " to the division sign " ÷ ". **Examples:**

6 = 25% of what number?
6 ÷ 25% "Change = to ÷."
6 ÷ .25 "Change % to decimal."

$.25\overline{)6.00}$ = .24

12 = 80% of what number?
12 ÷ 80% "Change = to ÷."
12 ÷ .8 "Change % to decimal."

$.8\overline{)12.0}$ = 15.

*Be careful to move decimal points properly.

Solve each of the following.

S1. 5 = 25% of what?

S2. 6 is 20% of what?

1. 12 = 25% of what?
2. 32 = 40% of what?
3. 5 is 20% of what?
4. 3 = 75% of what?
5. 12 is 80% of what?
6. 8 = 40% of what?
7. 15 is 25% of what?
8. 15 is 20% of what?
9. 9 is 20% of what?
10. 25 is 20% of what?

1.

2.

3.

4.

5.

6.

7.

8.

9.

10.

Score

Problem Solving

Bill took a test with 40 problems and got 36 of them correct. What percent of the problems did he get correct?

Review Exercises

1. Find 12% of 60.
2. Find 30% of 80.
3. 3 is what percent of 12?
4. 6 = what percent of 30?
5. 4 = 25% of what?
6. 5 is 20% of what?

Helpful Hints

Use what you have learned to solve the following problems. Use the following order.

1. Change = to ÷
2. Change % to decimal
3. Divide

* Be careful to move decimal points properly.
* "Is" means =.

Solve each of the following.

S1. 30 = 15% of what?

S2. 8 is 20% of what?

1. 5 = 25% of what?
2. 20 = 40% of what?
3. 3 is 5% of what?
4. 15 is 30% of what?
5. 10 is 40% of what?
6. 10 = 4% of what?
7. 7 is 20% of what?
8. 9 is 25% of what?
9. 16 = 20% of what?
10. 3 = 2% of what?

1.
2.
3.
4.
5.
6.
7.
8.
9.
10.
Score

Problem Solving

A bakery baked 250 cakes and sold 90% of them.
How many cakes were sold?

Review Exercises

1. Find 6% of 200.

2. 12 is what % of 48?

3. 3 = what 20% of what?

4. Change $\frac{12}{15}$ to a decimal.

5. Change 7.009 to a mixed numeral.

6. Change $6\frac{17}{1,000}$ to a decimal.

Helpful Hints

Use what you have learned to solve the following problems.

Examples:

5 people got A's on a test. This is 20% of the class. How many people are in the class?

5 = 20% of what number?
5 ÷ 20%
5 ÷ .20

$$.2\overline{)5.0} = 25.$$

There are 25 in the class.

200 students at a school are 7th graders. If this is 25% of the total students in the school, how many students are there in the school?

200 = 25% of what number?
200 ÷ 25%
200 ÷ .25

$$.25\overline{)200.00} = 800.$$

There are 800 students in the school.

S1. A team won 3 games. If this is 20% of the total games played, how many games have they played?

S2. Lucy deposited 150 dollars of her earnings into a savings account. If this was 25% of her earnings, how much did she earn?

1. James has 24 USA stamps in his collection. If that is 20% of his collection, how many stamps are in his collection?

2. A player made 9 shots. This was 75% of her total shots taken. How many shots did the player take.

3. 200 people eat cafeteria food at a school. If this is 40% of the school, how many students are there in the school?

4. 16 = 20% of what?

5. A man spent 8 dollars which was 5% of his earnings. What were his earnings?

6. A farmer sold 25 cows which was 20% of his herd. How many cows were in his herd?

7. 7 is 5% of what?

8. Ted has finished 3 problems on a test. If this is 15% of the problems, how many problems are on the test.

9. 12 players made the team. If this was 15% of all those who tried out, how many tried out for the team?

10. Sophia got 24 problems correct on a test. Her score was 80%. How many problems were on the test?

Answers
1.
2.
3.
4.
5.
6.
7.
8.
9.
10.
Score

Problem Solving

Anna bought groceries that cost $63.72. If she paid with four twenty-dollar bills, what is her change?

Review Exercises

1. 7.567 + 85 + .376 =

2. 3.19 - 1.776 =

3. $\begin{array}{r} .616 \\ \times\ .6 \\ \hline \end{array}$

4. 1,000 x 4.5 =

5. $.3\overline{)\,.027}$

6. $.005\overline{)\,16}$

Helpful Hints

Use what you have learned to solve the following problems.
* If necessary, refer to the examples on the previous page.

S1. If you get 35 questions right on a test, and this is 70% of the questions, how many questions are on the test?

S2. 20 people passed a test. This was 16% of those who took the test. How many took the test?

1. There are 3 girls in a class. If this is 20% of the class, how many are in the class?

2. A pitcher threw 9 strikes. If this was 75% of the total pitches thrown, how many pitches were thrown?

3. 25 = 20% of what?

4. There are 8 red marbles in a bag. If this is 40% of all the marbles, how many marbles are in the bag?

5. 15 students in a class were receiving awards. If this is 20% of the class, how many are in the class?

6. 6 is 40% of what?

7. Eva tipped a waiter 6 dollars. This was 15% of the bill. How much was the bill?

8. Mr. Pena paid $6,000 in taxes last year. If this was 25% of his earnings, what were his earnings?

9. Robert has saved 40 dollars. If this is 20% of the cost of a bike that he wants, what is the price of the bike?

10. 4 = 80% of what?

1.
2.
3.
4.
5.
6.
7.
8.
9.
10.
Score

Problem Solving

A man bought a CD player for $15.50. If state tax is 8%, what is the total cost of the CD player with tax included?

Review Exercises

1. $7 \overline{\smash{)}.056}$

2. $1.5 \overline{\smash{)}1.35}$

3. Change $\frac{5}{8}$ to a decimal.

4. $2.13 \times .05$

5. $6.5 + 7 + 2.23 =$

6. 5% of 30 =

Helpful Hints

Use what you have learned to solve the following problems. **Examples:**

Find 12% of 50.
.12 x 50

$$\begin{array}{r} 50 \\ \times .12 \\ \hline 100 \\ 50 \\ \hline 6.00 \end{array}$$

5 is what percent of 25?

$$\frac{5}{25} = \frac{1}{5}$$

$$5 \overline{\smash{)}1.00} = .20 = 20\%$$

6 is 25% of what?
6 ÷ 25%
6 ÷ .25

$$.25 \overline{\smash{)}6.00} = 24.$$

Solve each of the following.

S1. 4 is what percent of 20?

S2. 3 = 15% of what?

1. Find 20% of 210.

2. Find 6% of 350.

3. 15 is what percent of 60?

4. 5 is 20% of what?

5. 15 = 75% of what?

6. 30% of 200 =

7. 18 is what percent of 24?

8. Find 25% of 64.

9. 3 is 5% of what?

10. 16 is what percent of 80?

1.
2.
3.
4.
5.
6.
7.
8.
9.
10.
Score

Problem Solving

Izzy took a test with 50 problems and got 80% correct.
How many problems did he miss?

Review Exercises

1. $\begin{array}{r} 120 \\ \times\ .06 \\ \hline \end{array}$

2. $\begin{array}{r} 2.45 \\ \times\ .7 \\ \hline \end{array}$

3. $\frac{5}{8} \times 16 =$

4. $5\overline{)1.3}$

5. $8\overline{)2}$

6. $.05\overline{).013}$

Helpful Hints

Use what you have learned to solve the following problems.
* Refer to the examples on previous pages if necessary.

Solve each of the following.

S1. Find 80% of 360.

S2. 7 is 20% of what?

1. 3 is what % of 60?

2. Find 8% of 320.

3. 20 is 25% of what?

4. Find 40% of 60.

5. 12 = 50% of what?

6. 60 = what % of 80?

7. 3 = 50% of what?

8. Find 100% of 320.

9. 4 what % of 20?

10. Find 50% of 60.

1.
2.
3.
4.
5.
6.
7.
8.
9.
10.
Score

Problem Solving

Amy took a spelling test with 12 words on it. If she spelled 9 of the words correctly, what percent of the words did she spell correctly?

Review Exercises

1. Change $\frac{72}{100,000}$ to a decimal.

2. Change 2.0019 to a mixed numeral.

3. Change $\frac{9}{15}$ to a percent.

4. $\frac{3}{5} \times 25 =$

5. $8\overline{).168}$

6. $.3\overline{)2.4}$

Helpful Hints

Use what you have learned to solve the following problems.

Examples:

A man earns $300 and spends 40% of it. How much does he spend?

Find 40% of 300
.4 x 300

300
x .4
120

He spends $120.

In a class of 25 students 15 are girls. What % are girls?

15 = what % of 25?

$\frac{15}{25} = \frac{3}{5}$

$5\overline{)3.00}$ = .60 = 60%

60% are girls.

Five students got A's on a test. This is 20% of the class. How many are in the class?

5 = 20% of what?
5 ÷ 20%
5 ÷ .2

$.2\overline{)5.0}$ = 25.

25 are in the class.

S1. On a test with 25 questions, Al got 80% correct. How many questions did he get correct?

S2. A player took 12 shots and made 9. What percent did the player make?

1. A girl spent $5. This was 20% of her earnings. How much were her earnings?

2. Buying an $8,000 car requires a 20% down payment. How much is the down payment?

3. 3 = 10% of what?

4. A team played 20 games and won 18. What % did they lose?

5. A farmer sold 50 cows. If this was 20% of his herd, how many cows were in his herd?

6. 20 = 80% of what?

7. Paul wants a bike that costs $400. If he has saved 60% of this amount, how much has he saved?

8. There are 400 student in a school. 55% are girls. How many boys are there?

9. 12 is what % of 60?

10. Kelly earned 300 dollars and put 70% of it into the bank. How much did she put into the bank?

1.
2.
3.
4.
5.
6.
7.
8.
9.
10.
Score

Problem Solving

Nick's monthly income is $4,800. What is his annual income? (Hint: How many months are in a year?)

Review Exercises

1. $7.68 + 19.7 + 5.364 =$

2. $\begin{array}{r} 7.123 \\ -\ 4.765 \\ \hline \end{array}$

3. $\begin{array}{r} 3.14 \\ \times\ \ 7 \\ \hline \end{array}$

4. $\begin{array}{r} .208 \\ \times\ .06 \\ \hline \end{array}$

5. $3\overline{)1.44}$

6. $.15\overline{)1.215}$

Helpful Hints

Use what you have learned to solve the following problems.
* Refer to the examples on the previous page if necessary.

S1. Find 20% of 150.

S2. 6 is 20% of what?

1. 8 is what % of 40?

2. Change $\frac{18}{20}$ to a percent.

3. A school has 600 students. If 5% are absent, how many student are absent?

4. A quarterback threw 24 passes and 75% were caught. How many were caught?

5. Riley has 250 marbles in his collection. If 50 of them are red, what percent of them are red?

6. A team played 60 games and won 45 of them. What % did they win?

7. There are 50 sixth graders in a school. This is 20% of the school. How many students are in the school total?

8. A coat is on sale for $20. This us 80% of the regular price. What is the regular price?

9. Steve has finished $\frac{3}{5}$ of his test. What percent of the test has he finished?

10. Alex wants to buy a computer priced at $640. If sales tax is 8% what is the total cost of the computer?

1.
2.
3.
4.
5.
6.
7.
8.
9.
10.
Score

Problem Solving

Ann took five tests and scored a total of 485 points. What was her average?

Final Review of Percents

Change numbers 1 - 5 to a percent.

1. $\frac{19}{100} =$ 2. $\frac{7}{100} =$ 3. $\frac{9}{10} =$

4. $.27 =$ 5. $.3 =$

Change numbers 6 - 8 to a decimal and a fraction reduced to the lowest terms.

6. 3 % = . = ___ 7. 16 % = . = ___

8. 90 % = . = ___

Solve the following problems. Label the word problem answers.

9. Find 6% of 280.

10. Find 70% of 450.

11. Find 24% of 400.

12. 3 is what % of 15?

13. 24 = what % of 60?

14. 5 = 20% of what?

15. 6 is 5% of what?

16. Change $\frac{27}{36}$ to a percent.

17. Of the 300 students in a school, 40% are girls. How many girls are there in the school?

18. Cloe took a test with 35 questions and got 28 correct. What percent did she get correct?

19. Shawn has finished 18 questions on a test. This is 75% of the test. How many questions are on the test?

20. A team played 65 games and won 80% of them. How many games did the team lose?

1.
2.
3.
4.
5.
6.
7.
8.
9.
10.
11.
12.
13.
14.
15.
16.
17.
18.
19.
20.
Score

Final Test of Percents

Change numbers 1 - 5 to a percent.

1. $\frac{7}{10} =$ 2. $\frac{7}{100} =$ 3. $\frac{1}{10} =$

4. $.05 =$ 5. $.5 =$

Change numbers 6 - 8 to a decimal and a fraction reduced to the lowest terms.

6. 70 % = . = ___ 7. 2 % = . = ___

8. 15 % = . = ___

Solve the following problems, label the word problem answers.

9. Find 40% of 550.

10. Find 25% of 400.

11. Find 3% of 180.

12. 6 = 15% of what?

13. 15 is what % of 50?

14. 20 is 80% of what?

15. Find 8% of 210.

16. 24 is 25% of what?

17. A woman had $4,000 in her savings account. She withdrew 30% of it. How much did she withdraw?

18. On a baseball team there are 6 pitchers. If this is 15% of the team, how many players are on the team?

19. A man must pay a sales tax of 8% when purchasing a car. If the price of the car is $24,000, how much is the sales tax?

20. Rosita had 20 dollars. If she spent 16 dollars, what % of her money did she spend?

1.
2.
3.
4.
5.
6.
7.
8.
9.
10.
11.
12.
13.
14.
15.
16.
17.
18.
19.
20.
Score

Solutions

PAGE 4
Review Exercises
1. 97
2. 215
3. 451
4. 620
5. 109
6. 624

S1. three and seven tenths
S2. twelve and ninteen thousandths

1. eighty-seven hundredths
2. five and six thousandths
3. one hundred fifteen and seven tenths
4. seventy-eight and seven hundredths
5. six and three thousand nine hundred twelve ten-thousandths
6. eighty-five thousandths
7. seven and thirty-six hundredths
8. nine and two thousandths
9. sixty-one hundredths
10. two and three hundred thirty-three thousandths

Problem Solving: $180

PAGE 5
Review Exercises:
1. 1,408
2. 749
3. 523
4. 115
5. 224
6. 1,260

S1. three and six thousandths
S2. one hundred seventy-six ten-thousandths

1. eight tenths
2. three and five ten-thousandths
3. seventy-six and eight tenths
4. seven and eight thousandths
5. five and one hundred thirty-eight thousandths
6. fifteen thousandths
7. five and eighty-two hundredths
8. four and three hundredths
9. eighty-six hundredths
10. four and two hundred twenty-four thousandths

Problem Solving: 125 students

PAGE 6
Review Exercises:
1. 779
2. two and seven thousandths
3. 6,382
4. forty-two and sixteen thousandths
5. 327
6. nineteen thousandths

S1. 5.03
S2. 436.11

1. 7.4
2. 22.015
3. .0352
4. 74.043
5. .00005
6. .000016
7. 9.045
8. 20.0033
9. 86.9
10. 86.000009

Problem Solving: 360 crayons

PAGE 7
Review Exercises:
1. two and nine hundredths
2. two and nine thousandths
3. 2.04
4. 682
5. .015
6. 13,182

S1. 7.062
S2. .02009

1. 8.09
2. 12.00041
3. .0049
4. 97.000513
5. .048
6. 52.8
7. 5.496
8. 3.005
9. 12.00033
10. 116.05

Problem Solving: 24 students

PAGE 8
Review Exercises:
1. two and seven hundredths
2. seven and seventeen thousandths
3. 7.006
4. .000032
5. seventeen ten-thousandths
6. 5.0011

S1. 5.6
S2. 8.009

1. 21.16
2. .16
3. 14.017
4. 119.00016
5. .0021
6. .00196
7. 4.032
8. 3.324
9. 4.000017
10. .0019

Problem Solving: 450 mph

PAGE 9
Review Exercises:
1. 10.014
2. two and nine hundredths
3. .065
4. 10.00015
5. 107
6. 564

S1. 9.017
S2. 9.00017

1. 42.196
2. .072
3. 48.008
4. 16.00195
5. .016
6. 16.000119
7. 4.0038
8. 3.0176
9. .071
10. 6.0053

Problem Solving: 384 miles

Solutions

PAGE 10

Review Exercises:

1. 9.07
2. 1.00135
3. .00017
4. twenty-one thousandths
5. three and nineteen hundreths
6. 6.0013

S1. $3\frac{5}{100}$

S2. $16\frac{17}{1,000}$

1. $45\frac{19}{10,000}$
2. $\frac{5}{100,000}$
3. $7\frac{16}{1,000,000}$
4. $7\frac{196}{1,000}$
5. $79\frac{6}{10}$
6. $\frac{7,632}{100,000}$
7. $14\frac{7}{1,000,000}$
8. $16\frac{24}{1,000}$
9. $17\frac{145}{1,000,000}$
10. $\frac{96}{100,000}$

Problem Solving: $274

PAGE 11

Review Exercises:

1. 11.06
2. $\frac{6}{1,000}$
3. .019
4. $7\frac{6}{1,000}$
5. $6\frac{3}{10}$
6. .072

S1. $\frac{16}{10,000}$

S2. $9\frac{125}{100,000}$

1. $7\frac{9}{100,000}$
2. $\frac{16}{1,000}$
3. $7\frac{29}{100}$
4. $6\frac{2}{100,000}$
5. $87\frac{3}{10}$
6. $5\frac{72}{10,000}$
7. $15\frac{6}{1,000,000}$
8. $42\frac{1}{10}$
9. $163\frac{137}{10,000}$
10. $\frac{11,234}{100,000}$

Problem Solving: 6 boxes, 4 left

PAGE 12

Review Exercises:

1. $7\frac{16}{10,000}$
2. 7.015
3. 5.6
4. six and thirteen thousandths
5. three and seven thousandths
6. $72\frac{12}{10,000}$

S1. <

S2. >

1. <
2. >
3. <
4. >
5. >
6. <
7. <
8. >
9. <
10. <

Problem Solving: 8 gallons, $24

PAGE 13

Review Exercises:

1. $\frac{129}{10,000}$
2. seven and ninety-two thousandths
3. one hundred thirty-five ten thousandths
4. 17.006
5. 15.0071
6. $3\frac{96}{1,000}$

S1. <

S2. <

1. <
2. >
3. >
4. >
5. <
6. <
7. >
8. >
9. >
10. <

Problem Solving: $181

PAGE 14

Review Exercises:

1. $52\frac{623}{1,000}$
2. 75.009
3. 4,368
4. 50.019
5. six and twenty-three hundredths
6. 6,784

S1. 18.82

S2. 8.84

1. 41.353
2. 14.431
3. 29.673
4. 1.7
5. 19.26
6. 146.983
7. 1.7
8. 25.044
9. 42.055
10. 31.09

Problem Solving: 18.62 inches

PAGE 15

Review Exercises:

1. seven and nineteen hundredths
2. 14.503
3. 7.0005
4. $7\frac{632}{10,000}$
5. nineteen thousandths
6. 1.9

S1. 25.73

S2. 13.58

1. 50.103
2. 18.542
3. 34.976
4. 1.75
5. 107.36
6. 103.306
7. 2.3
8. 86.415
9. 56.602
10. 25.55

Problem Solving: $15

Solutions

PAGE 16

Review Exercises:

1. 19.557
2. 14.06
3. 7.0125
4. 6 12/1,000
5. two and seven hundredths
6. seven and five thousandths

S1. 13.84
S2. 7.48

1. 5.894
2. 2.293
3. 1.16
4. 11.13
5. .053
6. 3.9577
7. 2.373
8. 3.684
9. 8.628
10. 26.765

Problem Solving: 2.7°

PAGE 17

Review Exercises:

1. 2.6
2. 12.17
3. seventy-two hundredths
4. seventy-two ten-thousandths
5. 75 6/10,000
6. 97.03

S1. 55.62
S2. 73.04

1. 4.003
2. 7.187
3. 3.478
4. 36.34
5. .193
6. 4.8884
7. 4.036
8. 87.976
9. 2.286
10. 57.335

Problem Solving: $279

PAGE 18

Review Exercises:

1. 17.79
2. 5.058
3. 17.36
4. 12.22
5. .075
6. two and fifty-eight thousandths

S1. 18.38
S2. 3.687

1. 24.164
2. 6.17
3. 16.57
4. 8.24
5. 2.08
6. 19.56
7. 2.2
8. 4.487
9. 10.396
10. 25.7

Problem Solving: 5.5 pounds

PAGE 19

Review Exercises:

1. .000007
2. seven millionths
3. 79.22
4. 6.0022
5. 4.64
6. 2.335

S1. 49.417
S2. 6.764

1. 17.19
2. 2.2
3. 26.73
4. 3.234
5. 93.106
6. 2.063
7. 2.1
8. 66.218
9. 9.33
10. 73.001

Problem Solving: 347.95 miles

PAGE 20

Review Exercises:

1. 170
2. 5,551
3. 11,052
4. 20.69
5. 4.563
6. 16.5

S1. 7.38
S2. 36.8

1. 1.929
2. 14.64
3. 6.88
4. 56.64
5. 22.4
6. 55.2
7. 328.09
8. 10.024
9. 29.93
10. 1.242

Problem Solving: 38 tons

PAGE 21

Review Exercises:

1. 12.78
2. 32.2
3. 36.065
4. .736
5. 4.733
6. 19.747

S1. 17.35
S2. 101.2

1. 2.892
2. 22.32
3. 12.42
4. 86.88
5. 67.2
6. 79.35
7. 56.992
8. 17.82
9. 570
10. 3.288

Problem Solving: $100.75

Solutions

PAGE 22

Review Exercises:

1. 15.587
2. 5.656
3. 5.75
4. 3.84
5. 5.63
6. 22.636

S1. 2.52
S2. 7.776

1. 11.52
2. .4598
3. .21186
4. .874
5. .1421
6. 9.7904
7. .0024
8. 19.04
9. 149.225
10. .0738

Problem Solving: $8.00

PAGE 23

Review Exercises:

1. 38.52
2. three and twenty-six ten-thousandths
3. 3 $^{7}/_{1,000}$
4. 2.011
5. sixteen hundred-thousandths
6. 1.058

S1. .2268
S2. 13.692

1. 24.38
2. .6897
3. .18744
4. .2442
5. 1.242
6. 1.18209
7. .0042
8. 2.752
9. 15.2334
10. .1016

Problem Solving: $87.50

PAGE 24

Review Exercises:

1. 6.42
2. 3.621
3. .000006
4. 16
5. 1.857
6. $17.70

S1. 32
S2. 7,390

1. 93.6
2. 72,600
3. 160
4. 736.2
5. 7,280
6. 70
7. 37.6
8. 390
9. 73.3
10. 76.3

Problem Solving: $18,500

PAGE 25

Review Exercises:

1. two and seven hundred sixty-three thousandths
2. .00007
3. 4.45
4. .00016
5. .3615
6. 53.8

S1. 370
S2. 5,360

1. 327
2. 56,700
3. 190
4. 736.4
5. 976
6. 750
7. 73
8. 387
9. 8,330
10. 84.2

Problem Solving: 2,500 pounds

PAGE 26

Review Exercises:

1. 139.2
2. 9
3. 3,250
4. 4.236
5. .000009
6. 21.573

S1. 2.394
S2. 15.228

1. 3.22
2. .464
3. .48508
4. 26
5. 3.156
6. 2,630
7. .0108
8. 3.575
9. 5.511
10. .04221

Problem Solving: $72

PAGE 27

Review Exercises:

1. 11.75
2. 4.766
3. 75.382
4. 12.24
5. 14.88
6. 2,365

S1. 25.56
S2. .3596

1. 96.6
2. 2.961
3. .00627
4. 360
5. 89.28
6. .000005
7. 138.072
8. 42,900
9. 34.72
10. .02898

Problem Solving: $13.52

Solutions

PAGE 28

Review Exercises:

1. 22 r2
2. 303
3. 111
4. 361 r21
5. 132
6. 203

S1. .44
S2. 1.8

1. 19.7
2. 3.21
3. .57
4. 3.7
5. 6.08
6. 40.9
7. .324
8. .16
9. 2.12
10. 2.04

Problem Solving: $1.95

PAGE 29

Review Exercises:

1. 4.4
2. .345
3. 1.63
4. 5.483
5. .864
6. 78.29

S1. .232
S2. 2.68

1. .51
2. 4.06
3. .56
4. 5.6
5. 2.304
6. .23
7. .63
8. .16
9. .123
10. 4.08

Problem Solving: 12.2 seconds

PAGE 30

Review Exercises:

1. 7.38
2. .301
3. 270
4. 38.8
5. 279.66
6. $2\ ^{1,762}/_{10,000}$

S1. .0027
S2. .019

1. .0007
2. .012
3. .056
4. .036
5. .043
6. .063
7. .023
8. .022
9. .003
10. .068

Problem Solving: $33

PAGE 31

Review Exercises:

1. two and three thousandths
2. 110.213
3. 66.987
4. 436.1
5. 13.3
6. .172

S1. .055
S2. .038

1. .0004
2. .002
3. .072
4. .046
5. .034
6. .054
7. .057
8. .044
9. .0008
10. .034

Problem Solving: $5.25

PAGE 32

Review Exercises:

1. 67.8
2. .02052
3. 360
4. .003
5. .011
6. 6.00015

S1. .34
S2. .06

1. .065
2. .62
3. 2.05
4. .15
5. .04
6. .04
7. .12
8. 1.575
9. .006
10. .418

Problem Solving: $10.13

PAGE 33

Review Exercises:

1. $1.59
2. $63.20
3. .065
4. 47.48
5. 18.518
6. .08

S1. .074
S2. 2.25

1. .085
2. .078
3. .06
4. .16
5. .638
6. .15
7. 1.45
8. .185
9. 1.35
10. .04

Problem Solving: $1.81

Solutions

PAGE 34

Review Exercises:

1. .095
2. .34
3. .0002
4. 5.0013
5. 135/100,000
6. 2.2

S1. 3.9
S2. .24

1. 8
2. 170
3. .42
4. 80
5. 5.4
6. 3.2
7. 3.7
8. 3.2
9. 57.5
10. 9.2

Problem Solving: $2.13

PAGE 35

Review Exercises:

1. 2.07
2. 5.91
3. 4.104
4. .65
5. .034
6. .43

S1. 3.125
S2. 508

1. 1.25
2. .03
3. 5.08
4. 230
5. 20
6. 6.5
7. .33
8. 50
9. 9.2
10. 2.4

Problem Solving: 55 tiles

PAGE 36

Review Exercises:

1. 1.6
2. 15
3. .34
4. 300
5. 10.1
6. 5.2

S1. .75
S2. .625

1. .6
2. .25
3. .4
4. .875
5. .55
6. .52
7. .625
8. .2
9. .2
10. .7

Problem Solving: $420

PAGE 37

Review Exercises:

1. 2.3
2. 29.09
3. .2898
4. 4.877
5. 4
6. 50

S1. .375
S2. .75

1. .8
2. .1875
3. .44
4. .15
5. .75
6. .2
7. .26
8. .7
9. .28
10. .4375

Problem Solving: $2.76

PAGE 38

Review Exercises:

1. 10.48
2. 9.84
3. 123.6
4. 22.12
5. .0132
6. .38

S1. .005
S2. 40

1. .76
2. .074
3. .025
4. 4.5
5. .24
6. 284
7. 33.1
8. 16.2
9. .4
10. .625

Problem Solving: $9.50

PAGE 39

Review Exercises:

1. 19.7
2. .16
3. .003
4. 1.575
5. 54
6. 3.2

S1. .625
S2. 70

1. .0078
2. .78
3. 7.3
4. 5
5. 21.25
6. 286
7. .23
8. .65
9. .75
10. .25

Problem Solving: $12,000

Solutions

PAGE 40

1. 16.336
2. 13.87
3. 34.5
4. 27.9
5. 5.548
6. 70.51
7. 19.04
8. 128.8
9. 1.274
10. 1.74038
11. 3,190
12. 320
13. 1.98
14. .158
15. .32
16. 300
17. .03
18. 2.4
19. .625
20. .5625

PAGE 41

1. 41.716
2. 16.22
3. 2.9
4. 71.28
5. 52.877
6. .084
7. 4.685
8. 84
9. 3.528
10. .14283
11. 410
12. 3,762
13. .79
14. .0064
15. .25
16. 6,000
17. 470
18. 2.12
19. .125
20. .875

PAGE 42

Review Exercises:
1. .17
2. .9
3. 7/100
4. 7/10
5. six and seven hundredths
6. ten and nine tenths

S1. 17%
S2. 90%
1. 6%
2. 99%
3. 30%
4. 64%
5. 67%
6. 1%
7. 70%
8. 14%
9. 80%
10. 62%

Problem Solving: 12 gallons, $39

PAGE 43

Review Exercises:
1. .017
2. 500
3. 340
4. 4.284
5. 69.27
6. 4.136

S1. 70%
S2. 3%
1. 19%
2. 87%
3. 60%
4. 63%
5. 19%
6. 2%
7. 48%
8. 14%
9. 50%
10. 98%

Problem Solving: $172

PAGE 44

Review Exercises:
1. 70%
2. 72%
3. 400
4. .625
5. .6
6. .000006

S1. 37%
S2. 70%
1. 93%
2. 2%
3. 20%
4. 9%
5. 60%
6. 66.5%
7. 89%
8. 60%
9. 33.4%
10. 80%

Problem Solving: 9.6 fluid ounces

PAGE 45

Review Exercises:
1. 3 196/10,000
2. twenty-one ten-thousandths
3. 7.0007
4. 3.03
5. 4
6. .22

S1. 9%
S2. 34.8%
1. 90%
2. 9%
3. 70%
4. 9.7%
5. 60%
6. .7%
7. 87%
8. 30%
9. 44.5%
10. 40%

Problem Solving: 120 students

Solutions

PAGE 46
Review Exercises:
1. 7.11
2. 2.22
3. 19.24
4. 3.19
5. 2.463
6. 60.387

S1. .2, 1/5
S2. .09, 9/100
1. .16, 4/25
2. .06, 3/50
3. .75, 3/4
4. .4, 2/5
5. .01, 1/100
6. .45, 9/20
7. .12, 3/25
8. .05, 1/20
9. .5, 1/2
10. .13, 13/100

Problem Solving: 1/4

PAGE 47
Review Exercises:
1. 63.5
2. 18.375
3. .00024
4. .075
5. .202
6. 2.6

S1. .5, 1/2
S2. .05, 1/20
1. .08, 2/25
2. .8, 4/5
3. .24, 6/25
4. .11, 11/100
5. .02, 1/50
6. .7, 7/10
7. .09, 9/100
8. .9, 9/10
9. .17, 17/100
10. .14, 7/50

Problem Solving: $880

PAGE 48
Review Exercises:
1. .8
2. .07
3. 1/4
4. 109.2
5. 128
6. 18

S1. 17.5
S2. 150
1. 4.32
2. 51
3. 15
4. 112.5
5. 32
6. 80
7. 10
8. 216
9. 112.5
10. 13.2

Problem Solving:
160 miles per hour

PAGE 49
Review Exercises:
1. .4
2. .4
3. .65
4. 40.5
5. .375
6. three and sixteen ten-thousandths

S1. 24
S2. 240
1. 3.2
2. 385
3. 38.5
4. 50
5. 50
6. 186
7. 11
8. 240
9. 120
10. 6

Problem Solving: 84

PAGE 50
Review Exercises:
1. 7.2
2. 288
3. 10
4. .75
5. 1/2
6. 22.5

S1. 6 correct
S2. 30 absent
1. $84 in bank
2. $2,700 saved
3. 18 incorrect
4. $600 on food
5. 12 girls
6. $40,000 down payment
7. $5,600 value
8. $3.50, $53.50 total
9. 150 students
10. $1,500 on food & housing

Problem Solving: $297

PAGE 51
Review Exercises:
1. 48
2. 4.8
3. 180
4. .2
5. 2.3
6. 1.45

S1. 32 passed
S2. 18 girls
1. 540 cookies
2. 72 white marbles
3. 480 students
4. $3.60 saved
5. $3,600 down payment
6. $480 spent
7. $9.00 tip
8. $480 for rent & car
9. 30 did not pass
10. 360 boxes, 240 girls

Problem Solving: $7.50

Solutions

PAGE 52
Review Exercises:
1. 4.9
2. 7.2
3. 7.2
4. 8.44
5. 7.009
6. .09

S1. 20%
S2. 80%
1. 60%
2. 50%
3. 10%
4. 75%
5. 75%
6. 60%
7. 25%
8. 80%
9. 75%
10. 20%

Problem Solving: 144 people

PAGE 53
Review Exercises:
1. 93.29
2. 14.85
3. 1.519
4. .8
5. 11.1
6. .21

S1. 80%
S2. 75%
1. 50%
2. 70%
3. 25%
4. 20%
5. 25%
6. 25%
7. 62.5%
8. 50%
9. 25%
10. 37.5%

Problem Solving: 75%

PAGE 54
Review Exercises:
1. 46.5
2. 24
3. .75
4. 50%
5. 135
6. 600

S1. 25%
S2. 75%
1. 25%
2. 80%
3. 50%
4. 90%
5. 60%
6. 75%
7. 75%
8. 75%
9. 80%
10. 95%

Problem Solving: 320 cows

PAGE 55
Review Exercises:
1. 7.28
2. 3/1,000
3. 24
4. .9
5. 100
6. 3.328

S1. 75%
S2. 20%
1. 40%
2. 75%
3. 90%
4. 45%
5. 25%
6. 75%
7. 75%
8. 60%
9. 90%
10. 52%

Problem Solving: 36 correct

PAGE 56
Review Exercises:
1. 112
2. 3
3. 60%
4. 60%
5. 47.1%
6. 72.00601

S1. 60% correct
S2. 60% girls
1. 80% correct
2. 25% into savings
3. 25% lost
4. 60% caught
5. 90 percent
6. 20% absent
7. 48% won
8. 40% are sixth graders
9. 98% correct
10. 75% strikes

Problem Solving: $453.60

PAGE 57
Review Exercises:
1. 55%
2. 70%
3. 8
4. 20
5. 500
6. 500

S1. 75% did not get A's
S2. missed 25%
1. 80% won
2. 20% lost
3. 60% boys
4. 60% into savings
5. 60% take bus
6. 40% are goldfish
7. 25% missed
8. 90% have a computer
9. 75% completed
10. 30% have a pet

Problem Solving: $50.15

PAGE 58
Review Exercises:
1. 3.2
2. 32
3. 75%
4. 90%
5. 50.04
6. 200

S1. 20
S2. 30

1. 48
2. 80
3. 25
4. 4
5. 15
6. 20
7. 60
8. 75
9. 45
10. 125

Problem Solving: 90%

PAGE 59
Review Exercises:
1. 7.2
2. 24
3. 25%
4. 20%
5. 16
6. 25

S1. 200
S2. 40

1. 20
2. 50
3. 60
4. 50
5. 25
6. 250
7. 35
8. 36
9. 80
10. 150

Problem Solving: 225 cakes

PAGE 60
Review Exercises:
1. 12
2. 25%
3. 15
4. .8
5. $7\ ^{9}/_{1,000}$
6. 6.017

S1. 15 games
S2. $600 earned

1. 120 stamps
2. 12 shots taken
3. 500 students in school
4. 80
5. $160 earned
6. 125 cows in herd
7. 140
8. 20 problems on test
9. 80 tried out
10. 30 problems on test

Problem Solving: $16.28

PAGE 61
Review Exercises:
1. 92.943
2. 1.414
3. .3696
4. 4,500
5. .09
6. 3,200

S1. 50 questions
S2. 125 took test

1. 15 in class
2. 12 pitches thrown
3. 125
4. 20 marbles in bag
5. 75 in class
6. 15
7. $40 bill
8. $24,000 earnings
9. $200 for bike
10. 5

Problem Solving: $16.74

PAGE 62
Review Exercises:
1. .008
2. .9
3. .625
4. .1065
5. 15.73
6. 1.5

S1. 20%
S2. 20

1. 42
2. 21
3. 25%
4. 25
5. 20
6. 60
7. 75%
8. 16
9. 60
10. 20%

Problem Solving: 10 problems

PAGE 63
Review Exercises:
1. 7.2
2. 1.715
3. 10
4. .26
5. .25
6. .26

S1. 288
S2. 35

1. 5%
2. 25.6
3. 80
4. 24
5. 24
6. 75%
7. 6
8. 320
9. 20%
10. 30

Problem Solving: 75%

Solutions

PAGE 64
Review Exercises:
1. .00072
2. 2 19/10,000
3. 60%
4. 15
5. .021
6. 8

S1. 20 correct
S2. 75% made shots

1. earnings $25
2. $1,600 down payment
3. 30
4. 10% lost
5. 250 in herd
6. 25
7. $240 saved
8. 180 boys
9. 20%
10. $210 into bank

Problem Solving: $57,600

PAGE 65
Review Exercises:
1. 32.744
2. 2.358
3. 21.98
4. .01248
5. .48
6. 8.1

S1. 30
S2. 30

1. 20%
2. 90%
3. 30 absent
4. 18 caught
5. 20% are red
6. 75%
7. 250 in school
8. $25 regular price
9. 60% of test
10. $691.20 total

Problem Solving: 97

PAGE 66
1. 19%
2. 7%
3. 90%
4. 27%
5. 30%
6. .03, 3/100
7. .16, 4/25
8. .9, 9/10
9. 16.8
10. 315
11. 96
12. 20%
13. 40%
14. 25
15. 120
16. 75%
17. 120 girls
18. 80% correct
19. 24 on test
20. lost 13 games

PAGE 67
1. 70%
2. 7%
3. 10%
4. 5%
5. 50%
6. .7, 7/10
7. .02, 1/50
8. .15, 3/20
9. 220
10. 100
11. 5.4
12. 40
13. 30%
14. 25
15. 16.8
16. 96
17. $1,200 withdrawn
18. 40 on team
19. $1,920 sales tax
20. 80% spent

Math Notes

Math Notes

Math Notes

Made in the USA
Middletown, DE
29 September 2020

20770549R00046